T0356181

RISE OF THE ZOMBIE BUGS

RISE OF THE ZOMBIE BUGS

THE SURPRISING SCIENCE OF PARASITIC MIND-CONTROL

MINDY WEISBERGER

JOHNS HOPKINS UNIVERSITY PRESS
BALTIMORE

Printed in the United States of America on acid-free paper
2 4 6 8 9 7 5 3 1

Johns Hopkins University Press
2715 North Charles Street
Baltimore, Maryland 21218
www.press.jhu.edu

Library of Congress Cataloging-in-Publication Data

Names: Weisberger, Mindy, 1967– author.
Title: Rise of the zombie bugs : the surprising science of parasitic mind-control / Mindy Weisberger.
Description: Baltimore : Johns Hopkins University Press, 2025. | Includes bibliographical references and index.
Identifiers: LCCN 2024033275 | ISBN 9781421451350 (hardcover) | ISBN 9781421451367 (ebook)
Subjects: LCSH: Parasitism. | Parasites. | Host-parasite relationships.
Classification: LCC QL757 .W38 2025 | DDC 595.717/527—dc23/eng/20241127
LC record available at https://lccn.loc.gov/2024033275

A catalog record for this book is available from the British Library.

Special discounts are available for bulk purchases of this book. For more information, please contact Special Sales at specialsales@jh.edu.

To Hens, Martin, and my parents

Contents

RISE OF THE ZOMBIE BUGS

INTRODUCTION

Puppet Masters, Puppets, and What Makes a "Zombie Bug"

"They're coming to get you, Barbra!"
—JOHNNY (RUSSELL STREINER), *Night of the Living Dead*

A hand punches through the dirt atop a grave, and a gore-spattered creature pulls itself from the soil. A young woman's bloodied corpse twitches and reanimates, launching itself teeth-first at the throat of the nearest bystander. Pop idol Michael Jackson turns toward his horrified date and reveals himself as a hollow-cheeked, dead-eyed ghoul before leading a troupe of the undead in the first steps of that iconic "Thriller" dance.

Of all the myriad monsters that lurch through our imaginations and fuel our nightmares, zombies are uniquely fascinating and horrible. They might still look like whoever they were in life, but that resemblance is superficial at best and is usually a gruesome distortion of their former selves. Often their bodies are decaying, battered, hideously maimed, and broken. More dreadful still is how zombification changes their behavior. A zombie is no longer in control of its actions; rather, an external agent has seized control and now pilots its host's mind and limbs, animating them in a grim imitation of life that extends beyond death.

As horrific as zombies might be, their popularity in recent years is undeniable. The long-running cable TV series *The Walking Dead*, based on the comic book series of the same name, was one of television's biggest success stories; its bleak vision of desperate survivors navigating a zombie apocalypse ran for more than a decade, spawned multiple spin-offs, and enthralled many millions of viewers. The HBO adaptation of the popular video game *The Last of Us*, featuring a fungus that zombifies its human hosts, debuted in January 2023 to critical acclaim and immediately captivated audiences, setting a new viewership record for the cable network. Classic novels have been reimagined with undead monsters, such as *Pride and Prejudice and Zombies*, penned by Seth Grahame-Smith who also credits the original book's author, Jane Austen (the book was later adapted for the screen). There are zombie action movies (*Army of the Dead*), zombie comedies (*Shaun of the Dead* and *The Dead Don't Die*), zombie fare for middle schoolers (the Disney Channel's *Z-O-M-B-I-E-S* movies), and zombie romantic comedies, also known as zom-rom-coms (*My Boyfriend's Back*, *Warm Bodies*, *Life After Beth*, and *Lisa Frankenstein*). Zombie narratives drive popular video games such as *The Last of Us*, *Resident Evil*, *Left 4 Dead*, and *Plants vs. Zombies*. There are zombified versions of Marvel's pantheon of superheroes—including Iron Man, the Hulk, Spider-Man, Wolverine, and Ms. Marvel—in the comics series *Marvel Zombies*.

Zombies have even infiltrated reality television. In August 2023, Netflix debuted *Zombieverse*, an unscripted zombie-themed show. Set in Seoul, South Korea, the program tasked a group of contestants with surviving a zombie apocalypse by evading costumed zombies roaming the city and escaping to an evacuation ship (if a player was "bitten," they became a zombie and were eliminated from the competition).

Our fascination with zombies is nothing new; for centuries, people around the world have imagined shambling creatures that are somewhere between living and dead, puppeted by external forces. The notion of powerful individuals controlling others' bodies and minds, through magical or medicinal means, originated hundreds of years ago in West and Central Africa and then traveled to the Americas with enslaved people. Tales of zombies further evolved in Haiti and were then embellished by people scattered by the African diaspora. Stories of zombification spread even more widely across the globe in the twentieth and twenty-first centuries—initially as tales told in novels and comic books, and then through movies and other types of narrative media.

Such pop culture monsters are firmly fixed in the realm of fiction, but there are true zombies that have been here all along. For millions of years, arthropod zombies skittered, scuttled, and stumbled through diverse habitats across the planet. And they still do so today, piloted by parasites that hijack their hosts' nervous systems, reshape and distort their bodies in gruesome ways, and compel unnatural behaviors that benefit the parasite and the parasite alone, usually leading to the untimely deaths of the creatures they infect. These real-world zombifiers are small—sometimes microscopic—but the manipulative power they wield over their unfortunate hosts is insidious and absolute. Search just about any habitat where terrestrial invertebrates dwell, and you'll likely uncover a zombie presence.

In southwestern Brazil, a carpenter ant turns away from its nestmates and treks alone, climbing upward on a leaf, where it clamps its mandibles down tight; when it expires, fungal filaments creep from its limbs and a towering stalk erupts from its head. A female soldier beetle in a Florida meadow dies clinging to a flower. Her abdomen swells and she spreads her wings wide, assuming an alluring pose that invites males to become intimately acquainted with the parasitic fungus that killed her. A tiny, glittering tropical wasp stabs a cockroach directly in its brain, injecting it with chemicals that compel the roach to placidly follow the wasp on a dead-end journey to a tomb, where the roach will be buried and then eaten alive by the wasp's young. In

Central Europe, another species of tiny wasp plunges her stinger into a spider's mouth, delivering a potent chemical injection; the addled arachnid then spins a cocoon web to protect the wasp mother's larva, which the spider will shield and defend until it becomes the youngster's dinner. Near a canal in the Netherlands, tiny worms grow inside a snail until they are large enough for their movements to become visible through the snail's skin. The infected snail abandons its typical haunts in the undergrowth and creeps out into the open, where it can easily be spotted by hungry birds. Meanwhile, in the mollusk's swollen eyestalks, the parasitic hitchhikers create pulsing displays of patterns and colors that resemble those of squirming caterpillars, enticing birds to swoop down and eat the snails. The next stage of the worms' life cycle can then play out in the birds' guts.

Ant, beetle, roach, spider, and snail are but a few examples of organisms all over the world that fall victim to parasites that manipulate a host's behavior for their own benefit, while also controlling and wreaking havoc on the hosts' bodies. By zombifying a host—forcing it to behave in a manner that runs contrary to its natural habits or overrides its normal survival instincts—these parasites gain an advantage in their own struggle for survival. (For the purposes of this book, "zombie bugs" refers to a range of invertebrates: arthropods such as insects, spiders, and millipedes, and one mollusk—snails. Entomologists would argue that the word "bugs" is only accurate when referring to a subset of insects with sucking mouthparts, but I'm leaning on the colloquial, broader interpretation of the word, which is more forgiving and commonly serves as a blanket term for all of the diverse minifauna that creep around our backyards, parks, gardens, and homes. Sorry, entomologists.)

A zombified host might reject its own preference for shaded ground and seek sunlight, climbing to heights that help its fungal controllers disperse their reproductive spores. A host may become overly stimulated and hypersexualized, attempting to mate with males and females alike to share its zombifying infection as widely as possible. It may zealously guard and defend the young of its zombifying master. While under a parasite's control, a host may abandon the safety of the leaf-covered forest floor for open ground where predators hunt. It may lose its fear of natural predators. In some instances, it will hurl itself into the nearest body of water to drown—all so that the puppetmaster pulling its strings can then pass from the host's body into the guts of a different host species, to reproduce.

Some brain-manipulating zombifiers are animals with brains of their own, such as wasps, flies, and even minuscule worms called nematodes that have simple, ring-shaped brains containing most of their body's neurons. But a zombie-maker doesn't need to have a brain in order to practice mind-control. Certain types of fungi, viruses, and unicellular organisms without brains (or even nervous systems) can still tinker with a host's neurochemistry and nudge an infected creature to do their bidding. Fungi have been honing this ability since at least the Eocene epoch. One remarkable fossil, approximately 48 million years old, preserves signs of fungal zombification: scars on a fossilized leaf made by the "death grip" of zombie ants that had climbed to their doom.[1]

Fungal zombifiers frequently end up engulfing their hosts' corpses, sometimes blanketing them entirely and sprouting multiple spore-bearing pillars and stalks. The sight is so ghastly that it's infiltrated pop culture zombie lore, with infected humans becoming fungus-sprouting monstrosities. In *The Last of Us*, humans are horribly transformed by a type of fungus referred to as "Cordyceps." As it happens, this is an actual fungus genus that has long been associated with parasitizing and manipulating insects (though many of the zombifying species that were originally placed in the *Cordyceps* genus have since been reclassified under the genus *Ophiocordyceps*).

Unlike *Ophiocordyceps*-infected ants, which depart their colonies and seek a lonely leaf or twig after zombification, fictional fungus-infected human zombies become deadly hunters of uninfected humans and are brutal when they catch their prey. The so-called Cordyceps infection turns them into ravening creatures with no memory of their former lives and no recognition of friends or loved ones. They attack anyone who's uninfected, and their bites transmit the spores so that the parasitic fungus can spread to more hosts.

While that bloodthirsty behavioral quirk doesn't manifest in actual ant zombies, the physical appearance of late-stage Cordyceps-infected humans in these sci-fi worlds—with webs of fungal threads and bulbous growths sprouting from their bodies and heads—is hauntingly similar to the documented demises of zombified ants and other *Ophiocordyceps* victims.

Fortunately, real-world zombifying fungi and most other mind-controlling parasites restrict their horrific physical transformations and mind-control to insects and other invertebrates. However, humans may be more vulnerable to certain types of zombification than you may expect. Some species of mind-altering microbes have evolved to wreak havoc on mammalian brains, prompting erratic behaviors that promote transmission of the virus to new hosts. The disease known as rabies, for example, caused by the virus RABV in the *Lyssavirus* genus and transmitted through bites from infected animals, leads to brain inflammation in humans and other mammals, such as dogs, foxes, bats, and raccoons. It promotes aggressive behavior, which helps spread the virus to more potential hosts.

Another microbe, the single-celled protozoan parasite *Toxoplasma gondii*, causes the disease toxoplasmosis. Cats are the parasite's definitive hosts, where it mates and lays its eggs, but *T. gondii* can lurk and survive in just about any species of bird or mammal—including humans. *T. gondii* is known to alter the behavior of rodent intermediate hosts such as mice and rats, eroding their fear of cats. This leads to highly risky behavior that increases the likelihood of the rodents being caught and eaten, serving the parasite's purpose by delivering it exactly where it needs to be in order to reproduce: inside a feline's guts.

In recent years, scientists have discovered that rodents aren't the only intermediate hosts that switch up their habits when infected by *T. gondii*. Wolves that carry the parasite become more aggressive and are more likely to emerge as pack leaders,[2] and infected hyena cubs are bolder around lions.[3] What's more, humans are also susceptible to *T. gondii*'s mind-controlling influence.

Toxoplasmosis is present in approximately 30 percent to 50 percent of all humans on Earth—that's about two billion people worldwide,[4] with around 40 million infected in the United States alone.[5] In many of those who harbor *T. gondii*, the disease is dormant and there are no symptoms; it's more dangerous for people with compromised immune symptoms or for pregnant people. However, a growing body of evidence hints that even in cases where an infected person displays no other symptoms, *T. gondii* can change how they behave.

The parasite survives by producing proteins that deflect immune attacks from host cells; by manipulating immune responses or generating persistent stress in the immune system, *T. gondii* may disrupt normal brain function. Investigations into toxoplasmosis infections in humans suggest an association with certain psychiatric conditions, such as anxiety disorders, bipolar disorder, and schizophrenia. *T. gondii* can also affect dopamine production, which may trigger increases in confidence or in behavior related to reward-seeking, such as generosity and conscientiousness toward others. Though questions persist regarding precisely how *T. gondii* affects its human hosts, it seems that some type of brain manipulation is part of the package.

While humans aren't in imminent danger of suffering the same fate as a fungus-studded ant or a mind-controlled cockroach, scientific study of how zombifiers control their zombie bugs isn't just about predicting or thwarting an imminent zombie apocalypse. Much can be learned from zombification—sometimes it can even benefit humans. Insect zombifiers can help control invasive agricultural arthropod pests in ecosystems where the pests have no natural predators; because zombifying organisms often are species-specific, they target a host species but leave native wildlife untouched. Scientists are also analyzing the mechanisms through which zombie-makers manipulate their hosts. Pinpointing these chemical pathways in immune and nervous systems could lead to the development of new immunosuppressants, antibacterial agents, and cancer-fighting therapies. Insights from the world of zombie bugs also fuel research into the neurochemistry of behavior and the evolution of host-parasite relationships. Such research can illuminate understanding of human susceptibility to behavior-altering pathogens and may uncover novel treatments for neurodegenerative disorders.

Speaking for myself—as someone who's long been a horror movie buff and horror fiction nerd—I suppose it was bound to happen that I'd be drawn to the bizarre world of zombifying parasites and their hapless hosts. I read my first description of fungus-zombified ants more than two decades ago while working in the Exhibition Department at the American Museum of Natural History; I was researching parasites for a video about evolution and was intrigued to learn that zombification actually existed in nature (well, who wouldn't be?). I can't deny that the horror fan in me is deeply fascinated by the sheer grossness of zombification as a survival strategy, with all of its gore and messiness. But there's an elegance to zombification, too, in the precision of the zombifiers' manipulation of their victims.

Over the years, my curiosity about zombies has only deepened (much like an incurable zombie infection). As a science journalist, I've written about many types of zombifiers; the more that I read and the more that I talk to scientists who study these creatures, the more I appreciate the complexity of zombifiers' tools and chemical weapons, and the evolutionary pathways that honed their zombie-making powers to target certain species over others. I've also come to recognize that while these organisms may dominate and control their hosts, they are themselves uniquely vulnerable. Despite their astounding abilities, these parasites can't reproduce without their victims. In many cases, zombifying parasites have specialized to the point where they can only reproduce in a single host species, giving them even slimmer odds of reproductive success.

Perhaps I also find some uneasy comfort in getting better acquainted with the harsh realities of actual zombies in the natural world. If pop culture blockbusters are going to continue their current trend of spotlighting zombies and speculating about genetic mutations that could trigger an out-of-control global zombie pandemic in humans, then I want to know as much about the zombifying process as possible—just in case.

There are still plenty of unanswered questions about how zombifiers control their hosts. And for every new clue that scientists uncover about how zombification works—and what that might mean for people—the bigger and more complex the puzzle grows, with still more missing pieces revealed. But I won't deny that I also can't look away from a zombified insect in the thrall of its master, as the parasite's grip strengthens and the host's will drains away, and it staggers off to wherever its zombifier wants it to go. From spore-stuffed cicadas to disco-eyed snails, I can't get enough of these real-world zombies. By the time you're done reading this book, I hope the same will be true for you (if you weren't already a zombie fan when you started).

1

ZOMBIFIERS

An Origin Story

"How do you kill something that's already dead?"
—FREDDY (THOM MATHEWS), *The Return of the Living Dead*

It's a lovely spring day in the town of Leiden in the Netherlands, with barely a cloud in the sky and a delicious warmth from the sun that sinks into your bones and seems to promise that winter is truly over and summer is just around the corner.

You wouldn't know it from the spot indoors where I'm standing.

There, the air is cool and slightly damp, with a faint chemical tang. Racks of industrial metal shelving and stacked drawers stand atop a concrete floor and extend in rows along the length of the room. There are no windows anywhere, but the lighting tubes overhead are bright enough that the contents of glass containers lining the shelves are clearly visible.

When I peer closer at some of them, I almost wish they weren't.

I don't consider myself to be especially squeamish, but the sight of animal specimens bleached and stiffened by chemicals in the liquid that preserves them, can be a little unnerving; they seem more like flabby rubber props that were molded for an episode of *Doctor Who*, rather than creatures that were once living. The organisms that I'm here to see are especially otherworldly looking. For the most part, they're sinuous and eyeless, pale and wrinkled. When they were alive, they lived inside other animals and probably made life quite miserable for them in the process.

I'm currently deep in the heart of the invertebrates fluid collection at the Naturalis Biodiversity Center, a natural history museum and research center about 25 miles (40 kilometers) south of Amsterdam. It's home to approximately 43 million objects and is one of the largest natural history collections in the world. The main building that's accessible to the public is spacious, open, and filled with natural light. Five floors of exhibits are connected by sweeping staircases and feature architectural flourishes that recall shapes found in the natural world.

Collection storage space, by comparison, is built for practicality rather than aesthetics. There, I'm surrounded on all sides by thousands of slides, jars, and vials, many brimming with parasites. Leading the way through this maze of meticulously cataloged preserved life is a bearded and bespectacled Dutch scientist named Hannco Bakker. Bakker is the collection manager of Naturalis's invertebrate collections, which includes approximately 5.8 million specimens (not counting insects, which are in the separate entomology collection), and he has kindly agreed to show me a few of them. The receptacles that he selects from drawers and shelves all contain parasites in varying sizes, their serpentine shapes mottled in shades of ecru, cream, and beige. Some jars hold multiple parasitic organisms taken from a single host, while a few of the larger jars contain solitary specimens.

Most of what Bakker is showing me are helminths: parasitic worms such as flukes, flatworms, tapeworms, and nematodes. Many of the parasites in the museum's collection came to live there in 2006, when approximately 5,600 specimens were donated by the former Department of Veterinary Parasitology at the University of Utrecht. From 1950 until around 2000, veterinary faculty there had collected and preserved thousands of parasitic organisms, building a collection of more than 18,000 specimens.

Most were helminths and arthropods that they found in domesticated and wild animals. A second donation from Utrecht brought the Naturalis total to about 7,000 vials and 1,100 microscopic slides of parasite specimens. And today, I'm getting to meet a few of them.

Here in one vial that Bakker holds up is *Ophidascaris baylisi*, a nematode that was extracted from the stomach of a reticulated python (*Python reticulatus*). In another vial are examples of *Ichthyocotylurus variegatus*, tiny flatworms that infect fish. There are hundreds of them floating in the preserving fluid; each is about the size of an unshelled sunflower seed.

Earlier in the day, Bakker had pulled one very special example to show me. Submerged in clear liquid, the specimen fills about three-quarters of the large jar and resembles a dense pile of long, deflated balloons that have lost most of their color. A label on one side of the glass identified the parasite as *Diphyllobothrium stemmacephalum*, a type of tapeworm, collected in 2007 from a porpoise's intestine. A handwritten note taped to the jar further clarified what we were looking at: *In deze pot bevindt zich slechts één lintworm*, it read (translated from Dutch: "This jar contains only one tapeworm"). According to the note, the tapeworm in question measures 9.6 meters (about 31 feet) long, roughly the length of an Olympic-size swimming pool.

But the parasites that brought me to Leiden were much smaller than that. I had to squint to see these very tiny flatworms inside their finger-size vials: *Leucochloridium paradoxum*, the two labels read. One vial's contents were collected in the Netherlands in 1981 from *Succinea putris*, a small species of land snail. The other vial's contents were also collected from the Netherlands but nearly two decades earlier and were pulled from the cloaca (rear orifice) of a bird—a Eurasian jackdaw (*Corvus monedula*). Those two hosts—a snail and a bird—are vastly different, yet both play an essential role in the parasite's life cycle. *Leucochloridium* matures in the body of a snail, but it can only reproduce inside a bird. To get there, it manipulates the snail's behavior, making it more active in exposed places during the daytime, so that the snail is more likely to be eaten by a bird. To further ensure that its snail will be seen by a hungry bird, *Leucochloridium* hijacks the snail's eyes, inserting itself into the eyestalk during the parasite's larval broodsac stage. As *Leucochloridium* pulses there, its shifting pattern and colors—visible through the snail's stretched skin—resemble the undulations of a caterpillar, a delectable snack for avian predators.

The sight of these disco-eyed zombie snails is not uncommon in this part of the Netherlands, and they are particularly abundant in late summer in places near water, Bakker tells me. He has often seen them on warm days in August and September near wooden park bridges in the town of Hoogvliet, a city district of Rotterdam.

"You can see them crawling everywhere," he says, "with these eyes flashing over and over."

Of all the parasites in the Naturalis collection, few directly manipulate host behavior like *Leucochloridium* does with snails. But to understand zombification, we need to first look at parasitism. All zombifiers are by definition parasitic. During at least one stage of their lives—and often for multiple life stages—they rely on a host for their survival. The term "parasite" was first used in the sixteenth century and can trace its origins to the ancient Greek *parasitos*: "one who eats at the table of another." Symbionts also live on or inside other creatures, but they differ from parasites by enjoying a mutually beneficial or at least a neutral relationship with their hosts. However, when a parasite is involved, the parasite is the only one that benefits from the arrangement.

Bakker's own research focuses on tiny marine mollusks in the Triphoridae family, which parasitize sponges; other types of sea snails parasitize sea stars, sea cucumbers, sea urchins, and corals. In many cases, one species of mollusk parasitizes just one host species; evolutionary pressure to specialize among potential hosts could explain why mollusks are one of the most diverse groups in marine environments, Bakker says.

The earliest direct fossil evidence of a parasitic relationship comes from the ocean and is about half a billion years old, preserved in tiny shelled marine animals called brachiopods. About 512 million years ago in an ocean that covered what is now southern China, soft-bodied worms built mineralized tubes on brachiopods' shells. Scientists from Northwest University in Xi'an, China, studied the fossils and suggested that the worms were kleptoparasites—food-stealing freeloaders that devoured floating particles that drifted toward their filter-feeding hosts. Hundreds of these brachiopod fossils—some parasitized and some untouched—were found at the site in China, enabling researchers to compare individual brachiopods and see how being parasitized might have affected their fitness. It turned out that parasitized brachiopods were smaller than their nonparasitized neighbors, hinting that the worms stole so much of their hosts' food that the loss stunted the brachiopods' growth.[1]

When the scientists published their results in *Nature Communications*, I contacted study coauthor Timothy Topper, a researcher at the Swedish Museum of Natural History in Stockholm, to ask about other signs of parasites in the fossil record. I was surprised to hear that there are lots of examples, and they're quite diverse: "from fleas on mammals to mites on insects, and even potentially single-celled parasites on *Tyrannosaurus rex*," Topper said at the time.[2]

Often, the evidence of parasitism is preserved as damage left behind, such as bumps in the shells of marine shrimps dating to 110 million years ago.[3] In some instances the parasites themselves are preserved as eggs in coprolites—nuggets of fossilized poop. Very occasionally, fossils turn up with evidence of a parasite embedded in the body of its host. The brachiopods from China are one example. Another is inside a chunk of Baltic amber dating to the Cretaceous period, found at a site in Russia. About 45 million to 55 million years ago, sticky tree sap flowed over a dead carpenter ant. As the sap hardened into amber, it enclosed not only the ant but also the stalks of a parasitic fungus that were sprouting from just behind the ant's head. Further compromising the dead ant's dignity was a mushroom cap poking from its rear end.[4]

A direct sign of ancient parasitism was even discovered inside the fossil of a long-necked titanosaur, found in the state of São Paulo in southern Brazil. Numerous lesions on the bones hinted that the sauropod suffered from inflammation so severe that infection would have eaten away at the animal's soft tissue all the way through to its skin and caused an abundance of open sores. This would have made the titanosaur's appearance quite gruesome, the scientific team surmised, so they nicknamed it "Zombie Dino" (a reconstruction by artist Hugo Cafasso shows the gory extent of the skin ulcerations, with gaping wounds visible on the dinosaur's body, neck, tail, and limbs). When the researchers sliced the fossil to look more closely at the sites of inflammation, they discovered something else preserved in the dinosaur's bones: over 70 microfossils of wormlike organisms that the researchers identified as blood parasites, and the cause of Zombie Dino's horrific infection.[5]

When humans evolved, parasites were right there alongside them. Our relationship with human-infesting parasites is ancient and long-standing—head lice in the genus *Pediculus* that target humans, for instance, are part of a group that has coevolved with primate hosts for approximately 25 million years.[6] The oldest mention of parasitic infection in historic records is the Ebers Papyrus from Egypt, dating to about 1500 BCE, and over subsequent centuries, parasites earned mentions in medical texts penned by physicians from Greece, China, India, and the Roman and Arab Empires. Archaeologists, too, have excavated plenty of evidence showing that the lives of people and their parasites have been closely intertwined for thousands of years. There are Bronze Age human coprolites full of parasite eggs in an ancient village in England known as "Britain's Pompeii"; intestinal parasites in mummies from China's Ming Dynasty; and parasitic protozoans in biblical-era toilets in Jerusalem. Worldwide, parasites' presence and influence is recorded in human history and preserved in fossils, remains, and artifacts.

To date, scientists have identified hundreds of species of parasites that infect people. From ectoparasites like ticks and lice that live on our skin and in our hair, to endoparasites like tapeworms that make themselves at home inside our bodies, our past and present are richly infused with parasites. Some, referred to as heirloom parasites, have been with *Homo sapiens* since our species first emerged in Africa and are a legacy of our primate ancestors. Others, known as souvenir parasites, were acquired and then shared through contact with other animals, plants, and ecosystems, after humans dispersed and spread around the world. Finding and tracking parasites across historical records and archaeological objects—especially toilets—reveals the impact of parasitic infection on large populations. The ebb and flow of parasites has fueled the spread of epidemics and brought armies to their knees, but has also led to positive outcomes like public sanitation reform.

Parasites have likely been around on the planet nearly as long as there has been life available to be parasitized, says biologist Kelly Weinersmith, a behavioral ecologist specializing in parasitic interactions, and an adjunct assistant professor in the Department of BioSciences at Rice University in Houston.

"My fellow parasitologists [and I], we like to joke that the first organism was free-living and the second organism that came on the scene figured out a way to exploit the first," she says. Weinersmith studies a peculiar parasite–host pairing, in which both parties are parasitic wasps. The crypt-keeper wasp (*Euderus set*) is a shimmering, iridescent jewel of an insect. It is a parasitoid, which means that its parasitic larvae typically kills its host. It's also a behavior-manipulator and engineers a gruesome demise in crypt gall wasps, in at least six different species.

Gall wasps are also parasites—of trees. They chemically induce the growth of galls; these knobby lumps sprout like tumors on leaf stems, and the wasps use them as "crypts" to raise their young. On the inside, galls have a chamber for a wasp egg. There's enough room for the larva to grow and develop, nourished by the gall's tissues. But sometimes an immature gall wasp ends up with an unexpected roommate, if a female crypt-keeper wasp happens to come across the nursery chamber and inserts her own egg inside it.

"We don't know yet if she's laying the egg inside of the host [the gall wasp] or just in the crypt with the host; we haven't been able to capture that part of the life cycle in the lab," says Weinersmith. While examining infected galls, the researchers at one point found a crypt-keeper wasp half in and half out of a gall wasp body. But they were uncertain if the parasitoid had gnawed its way out of the host, or was in the process of eating its way in.

Under normal circumstances, a gall wasp would transform peacefully into an adult while inside the gall, then chew its way out. Add a crypt-keeper to the mix, and the gall wasp is destined to catch only a fleeting glimpse of freedom. It will live long enough to make a small hole in the outer wall of its tomb, but the opening won't be large enough for the wasp to escape. Instead, it will end its life with its head wedged into the too-small gap, one eye peering up at the sky while the crypt-keeper larva that's still inside the chamber eats it alive. Once the host has been devoured to the exoskeleton, the crypt-keeper undergoes its own metamorphosis. As an adult wasp, it enters the world by bursting through the head of its deceased host.[7]

A key question that scientists ask when looking at what might be parasitic behavior manipulation of a host is whether or not the host's deviation from normal behavior benefits the parasite.

In the case of the crypt-keeper, Weinersmith and her colleagues suspected that the wasp larva needed its host not only as dinner but also as a way to breach the sealed gall and climb free. To do that, crypt-keepers would need to manipulate the gall wasp's hole-making. The researchers tested that hypothesis by attaching small pieces of bark over crypt openings while the crypt-keepers were still inside. They found that if the crypt-keeper had to punch through even a thin layer of bark on its own, "it was three times more likely to die trapped inside the crypt," Weinersmith says. The crypt-keeper's success, they hypothesized, depended on two things: keeping its host alive long enough for it to chew an opening and killing the host before it could gnaw a hole big enough to slip through. To that end, crypt-keepers likely evolved tools that enabled them to control the host wasp's normal hole-chewing behavior—though how that happens is still a mystery, says Weinersmith.

"I can imagine it ranging from tinkering with things like hormones, to targeting when they kill the host at a very particular time," she says.

Parasites as a group must be doing something right. Of the roughly 7.7 million known animal species, an estimated 40 percent are thought to be parasitic, and the strategy has evolved independently at least 223 times across the animal family tree.[8] Bakker's marine snails parasitize sponges, but there are also sponges that parasitize snails. Vertebrates that feed on living animals without killing them are a type of parasite; there are fish that nibble on the scales of living fish, and birds as well as bats that drink the blood of mammals and other birds (ironically, "vampiric" bird behavior is thought to have evolved from eating parasites off their host's skin and fur). Certain invertebrate groups, such as worms and arthropods, have evolved more than their fair share of parasites. By some estimates, helminths alone may include as many as 300,000 parasitic species.[9] And wasps leave worms in the dust when it comes to parasitism, with an estimated 1 million parasitoid wasp species worldwide.[10]

One bizarre group of insects went all-in on parasitism: strepsipterans, also known as twisted-wing parasites, in the order Strepsiptera. There are more than 600 species distributed worldwide, and all of them are parasitoids of other insects, including ants, bees, wasps, leafhoppers, and crickets. Strepsipterans are tiny, with most measuring just a few millimeters long. The group has been around for at least 65 million years, but don't feel bad if you've never spotted one; even entomologists struggle with that (according to the United Kingdom's Royal Entomological Society, strepsipterans are "hardly seen even by entomologists unless they make a special effort to study them").[11] These insects have evolved a deeply weird reproductive strategy—even for a parasite—which is just one of the reasons why they're extremely tricky to find, capture, and identify.

Both male and female strepsipterans are endoparasitic during their larval stage; they fool their host's immune system by hiding inside a protective sack made from the host insect's own epidermal cells. But while all males emerge as free-living adult insects, strepsipteran females in the Stylopidae family never leave their hosts and maintain a larva-like appearance even after metamorphosis; they mate, reproduce, and die while embedded in the host's body (only the exposed part of a female's body—her head—transforms from its larval state). An adult female strepsipteran has no legs and only rudimentary eyes and antennae; she's basically a bag of reproductive organs sticking partway out of the host, says biologist Jeyaraney Kathirithamby, who has studied strepsipterans at the University of Oxford since the mid-1970s (see figure 1).

In some species, females are parthenogenetic, reproducing without males. In other species, a male mates with a female that's still embedded in a host insect and poking headfirst out of its abdomen, emitting a pheromone that attracts her suitors. Adult

Figure 1 A row of three strepsipterans (black ovals) in the genus *Xenos*, poking out of the abdomen of a wasp (*Polistes dorsalis dorsalis*). *Courtesy of Sean McCann*

males live for just a few hours, so they waste no time inseminating as many females as possible before they drop dead, depositing their sperm inside a pouch in the eyeless, mouthless female's exposed head. After larvae hatch inside their mother, they exit the host and seek a new one.

To make things especially challenging for scientists who are trying to classify and study strepsipterans, males and females differ dramatically in more than their looks. Some species parasitize entirely different families of insects. "That's where the difficulty of identifying them comes into being, particularly in a group called Myrmecolacidae, where the males parasitize ants, and the females parasitize crickets and grasshoppers," says Kathirithamby. Matching up strepsipteran males and females of the same species in those cases is, understandably, next to impossible without a technique known as DNA barcoding, which compares short strands of genetic information.

Some studies of the group have suggested that strepsipterans manipulate their hosts, leading wasps to abandon their colonies or bees to change pollinating behavior. But because strepsipterans are so elusive and challenging to find and identify, despite decades of research, there are still many open questions about their parasitic habits, evolution, and ecology, let alone how they might be manipulating the insect that they parasitize, Kathirithamby adds.

"It's a very intriguing group," she says. "The more you look at it, the more there is to study."

*

Parasitism is a success story for numerous animal species, but it's not only animals that embrace the parasitic lifestyle. About 1 percent of flowering plants—roughly 4,000 species—are parasites.[12] Many types of fungi are parasitic, living off plants, animals, and even other fungi. Viruses are parasites by nature of their reproduction; they can only make more viruses by hijacking the cells of a host. Protozoa—single-celled microorganisms in the Protista kingdom—can also be parasitic, such as *Giardia*, which causes the disease giardiasis, and *Plasmodium*, the cause of malaria.

Along their evolutionary journey, some types of parasites picked up the ability to modify host neurochemistry and alter their behavior, to benefit the parasite's growth and improve its chances of making more parasites. For a parasitic fungus or virus, that might mean sending an infected host on a death march to an elevated location from which the parasite can more easily disperse to infect another wave of hosts. In parasitic worms such as the snail-infecting *Leucochloridium*, it means modifying the normally reclusive behavior of the intermediate host so that it's more likely to be found and eaten by the parasite's definitive host.

However, not all changes in host behavior are evidence of manipulation, says Shelley Adamo, a professor at Dalhousie

University in Halifax, Nova Scotia, who studies this relationship in insects. "The key thing about parasitic manipulation is it's an evolved response on the part of the parasite that has to somehow benefit the parasite," Adamo says. Sometimes, infected hosts' behavior changes are defensive, generated by their own immune systems. For example, an insect may react to infection by seeking a warm spot and basking, in order to drive out the parasite. "That's called behavioral fever," Adamo explains. "In many cases, it helps them get rid of pathogens." Other behavior changes may be a by-product of damage caused by the parasite nomming its hosts' body tissues—but the parasite doesn't necessarily benefit from the damage.

In the case of true behavior manipulation, "the parasite has been selected to secrete something that can alter the behavior of the host, or somehow manipulate nervous system function," says Adamo. "And then you can see how that affects behavior, and that behavior should benefit parasitic reproduction, or parasitic transmission, or something that is going to increase the fitness of the parasite."

Some manipulators achieve their goals by dosing their host with compounds that are recognizably mind-altering, such as the fungus *Massospora cicadina* that shapes cicada behavior with psilocybin and cathinone, a type of amphetamine. Certain types of jewel wasps deliver brain-changing chemicals by stabbing their victims directly in their brains. But in most cases, there's no obvious delivery system or chemical silver bullet for a behavioral change. Parasites may use multiple pathways to evade a host's immune responses and disrupt the central nervous system (and often the immune system) to cause abnormal behavior. One approach is for the parasite to chemically co-opt existing host systems, inducing the host to produce more or less of a certain hormone or other regulatory chemical, Adamo wrote in *Current Opinion in Insect Science*. Some behavior-manipulating viruses do this to caterpillars, deactivating the production of a hormone that initiates molting.[13]

But pinpointing these changes—and tracking them directly to the intervention of a parasite—isn't easy. "The more complicated the behavior, the less certain we are as to the trajectory of information and how that information is processed," says Adamo. What's more, insect brains have their own immune responses that can also generate behavioral changes, making it even trickier to unravel exactly what a parasite is doing.

Another hurdle in understanding manipulation is that some forms of unusual behavior are comparatively easy to observe; others, not so much. Disco-eyed snails, cocoon-guarding caterpillars, or leaf-gripping ants are a few dramatic examples. But not all are so obvious. In other cases, "the change in behavior could be something like the animal spending 10 percent more of its time mobile, than if it were uninfected," says parasitologist Robert Poulin, a professor of zoology at the University of Otago in New Zealand. "You cannot see that from a casual observation."

To date, there are thought to be several hundred associations of parasites and hosts that demonstrate at least some degree of control over host behavior, spanning all major phyla on the tree of life. This suggests that the strategy has evolved independently many, many times, explains Poulin. "Exactly how many times is a little bit difficult to determine at the moment because it has not been studied in a range of organisms. So it's difficult to know exactly whether it's a characteristic of an entire lineage or only some species in that lineage," Poulin says. "But it's certainly a strategy spread across all living organisms."

In 1931, Eloise B. Cram, a zoologist for the US Department of Agriculture, was one of the first to report host manipulation by a parasite. Her example was the nematode species *Tetrameres americana* affecting the behavior of their intermediate hosts—grasshoppers—and how that might help the parasite infiltrate the guts of its definitive bird host. Cram reported that nematode larvae grew in the leg muscles of grasshoppers and that infection made the insects less active, "a condition that would make them easy prey for food-seeking fowls in nature."[14]

Over the decades that followed, a growing number of researchers turned their attention to the parasite–host relationship, noting behavior changes that could represent parasitic manipulation through tissue invasion or neurochemistry—or both. In 1972, John C. Holmes and William M. Bethel, both researchers in the Department of Zoology at the University of Alberta in Canada, outlined multiple ways that hosts' habits reflected parasitic disruption, including disorientation, decreased stamina, and exposing themselves to potential predators.[15]

Parasitologist Janice Moore, now a professor emeritus of biology at Colorado State University, was among the first scientists to delve deeply into the intersection of parasites and host behaviors. "The phenomenon is not even rare," she reported in *Scientific American* in 1984. "One need only look in a lake, a field or a forest to find it."[16]

At the time, Moore was looking at parasitic thorny-headed worms that infect terrestrial isopods called pill bugs as intermediate hosts (to reproduce, the worms need to end up inside starlings or other songbirds). Healthy pill bugs hide under leaf litter, but infected bugs tended to wander in exposed areas where birds can easily spy them and snap them up, Moore found. Other species of thorny-headed worms that infected different arthropods as intermediate hosts also reproduced in vertebrates—on land and in water. Moore's review of prior studies noted that species across three classes of thorny-headed worms controlled host behavior in ways that paralleled what she documented in her own research, causing their hosts to be more active in open, well-lit areas than they normally would be. Such manipulation, she concluded, "may well be universal in the phylum."

A new perspective on the interplay between parasite and host arose in 1982: a biological concept called "extended phenotype," coined by British evolutionary biologist and author Richard Dawkins. A phenotype is all the observable traits in any individual. These can be inherited traits—an animal's hair texture or a flower's color—or traits shaped by environmental factors such as nutrition or disease. An extended phenotype refers to the expression of an organism's genes outside itself; for instance, when the genes of a parasite trigger changes in the phenotype of a host, affecting its appearance or behavior.

So-called zombie ants, which are parasitized and manipulated by fungus in the *Ophiocordyceps* genus, display this during the latter stages of infection, researchers wrote in *BMC Ecology*. When an ant staggers drunkenly away from its colony and climbs to its death, it is entirely under the fungus' control; it has no choice but to follow where the fungus leads it. "While the manipulated individual may look like an ant," the scientists reported, "it represents a fungal genome expressing fungal behavior through the body of an ant."[17]

How an animal behaves is coordinated through a complex array of electrical signals that zing through its brain and nervous system, with messages chemically relayed to neurons through hormones and by molecules called neurotransmitters; activating certain neurons triggers responses in the body. Over many decades of research, scientists have pieced together a detailed view—on the molecular and cellular level—of how nervous systems produce, transmit, and receive signals, and how that connects to behavior. The part that's less understood "is exactly how you wire up a behavior and modulate it," Adamo says. Even seemingly simple behaviors may require precise coordination of multiple signals within the central nervous system. And the more complex the behavior, "the less certain we are as to the trajectory of information and how that information is processed." Parasites that zombify their hosts may induce an immune system response; that in itself can trigger behavior changes. But the parasite may also produce other compounds—or induce the host to produce them—to change behavior even further.

"Teasing apart those two different mechanisms is really very, very hard," Adamo says. "It's not like we can't find changes in the brain due to these parasites that are changing behavior," Adamo adds. "It's that in the sea of changes, which ones are actually causally important for the change in behavior and how exactly do the parasites induce those particular changes. That's really the challenge for the field at the moment."

Because parasites have in some cases coevolved with their hosts for many millions of years, they've had plenty of time to tinker with host physiology and come up with molecular keys to unlock specific functions of the nervous system and the immune system. Some of those molecules have been identified, but multiple compounds may be involved, produced at different times in the host's and parasite's development. It's a level of control that researchers in the emerging field of neuroparasitology—an area of biology specializing in parasites that usurp hosts' neurochemistry—can only dream of reproducing. *Ophiocordyceps*, for example, "can change ant behavior in ways that scientists can't do in the lab," Weinersmith says. "The fungus 'knows' the ant nervous system better than we do."

For most parasites, the key to manipulation is likely many "small nudges," affecting multiple parts of a neurotransmitter system—or interconnected systems—rather than knocking a system out entirely, Adamo suggested. "It's these many small decreases that end up giving you these changes in behavior. Because we don't quite understand the circuitry, it's difficult to really spell out how that is happening—but I think that's what's happening," says Adamo. "By understanding these things, it might give us some insight into more subtle ways of causing more permanent changes in brains and behavior."

As some aspects of zombification are resolved, the riddles still keep coming. There are zombifiers that aren't close relatives but provoke similar behaviors in their hosts; do they use the same pathways to get those results? In closely related parasitic species, why is one species a manipulator, but its cousin isn't? And does the parasite's own microbiome play a role in host zombification?

"These are some of the cool questions on the horizon for the next few years," says Poulin, the University of Otago parasitologist. "There's still so many things to answer about this phenomenon."

*

When I emerged from the Naturalis Biodiversity Center into the late afternoon, I had parasites on my mind (not literally, thank goodness). As I walked across the campus toward the Leiden town center, I spotted a tiny snail slowly but determinedly crossing one of the footpaths. I gently lifted it up to move it to a safer patch of grass, but before setting it down I couldn't help myself; I had to take a closer look at its wee eyestalks. I could almost imagine that I saw them stretching toward me in surprise; the snail equivalent of raising its eyebrows. Though it appeared to be free of pulsating psychedelic-looking parasites, I still wondered if there was a parasite lurking somewhere inside the snail's body, biding its time until it was ready to zombify the unsuspecting mollusk. There was no way for me to tell for sure (at least, not through eye contact). I deposited the snail on the grass, and as I watched it inch away I thought of the question that Moore posed in 1995, in *BioScience:* "When you see a snail move through the intertidal or observe a terrestrial isopod crawl across the sidewalk, what organism is at the controls—the snail, the isopod, or a parasite?"[18]

2

FUNGUS AMONG US, PART I

Zombie Ants

"If you turn into a monster, is it still you inside?"
—SAM (KEIVONN WOODARD), *The Last of Us*

When HBO's dramatic series *The Last of Us* debuted in 2023, an episode early in the season quickly captured viewers' and critics' attention with a pivotal scene: The Zombie Kiss (be warned: there are spoilers ahead).

In a tense moment toward the end of episode 2, smuggler Tess (Anna Torv) revealed that she had been bitten by an Infected—one of the "zombies" of the series. Knowing that she was doomed, Tess persuaded her traveling companions Joel (Pedro Pascal) and Ellie (Bella Ramsey) to go on without her. Tess prepared an explosive trap, waiting to set it off until the zombified mob entered the building where she was hiding. Frenzied Infecteds soon broke through the door, one beelining for Tess as she fumbled to ignite her makeshift bomb. Just before the bomb exploded, the Infected leaned in close and opened its mouth . . . and squirming mycelia—networks of fungal threads—reached greedily toward Tess's petrified face.

That moment was uniquely horrifying—perhaps even more so than most zombie infection scenes—because fungi and their mechanisms for infection, behavior manipulation, and reproduction are so rarely seen in pop culture representations of zombies. (The show did take some dramatic liberties with portraying exactly *how* its zombifying fungus spread infection; fungi typically reproduce through aerial spore dispersal rather than "kissing" their prey with mycelia, but more on that later.)

While fungi-themed zombification may be a relatively new plot twist for zombie fiction, the genre had already been booming for at least a decade. Between 2000 and 2009, more than 450 zombie titles lurched onto screens worldwide, by IMDB's accounting: George A. Romero's *Land of the Dead,* Zack Snyder's *Dawn of the Dead* (a remake of Romero's 1978 classic), the dark comedy *Shaun of the Dead,* and the Will Smith–led *I Am Legend* are just a few examples. Across hundreds of zombie stories, the causes of zombification often went unexplained or were vaguely attributed to viral pathogens.

Even more zombie fare arrived with the 2010s, when the rise of streaming services released a tsunami of zombies—IMDB lists more than 950 zombie-themed series and movies produced between 2010 to 2020. That decade also introduced a zombifying pathogen that had previously been overlooked: parasitic fungus. It emerged spectacularly in the 2013 video game *The Last of Us,* and in the 2014 novel and subsequent film *The Girl with All the Gifts* written by M. R. Carey. (When *The Last of Us* was later adapted by HBO, the game's original writer and creative director, Neil Druckmann, was a cocreator and writer.)

Finally, fungus was getting its turn in the zombifying spotlight.

The fungus in *The Last of Us* (game and HBO series alike) is identified as "Cordyceps," and infection leads to a syndrome called Cordyceps brain infection, or CBI. *The Girl with All the Gifts* is more specific, naming its pathogenic fungus as a mutated version of the species *Ophiocordyceps unilateralis.* In both cases, the fungus infects and then mind-controls its victims, overriding their will and intellect and driving them to aggressively attack and spread the infection to anyone they see (or hear, or sense through movement). Eventually, as the parasitic pathogen grows and extends its reach throughout its hosts' bodies, it physically transforms them; their limbs, trunks, and heads sprout layers of fungal structures until they resemble monstrous aggregations of oyster mushrooms come to life.

Though these fungal zombies are fictional, their inspiration has its roots (or rather, mycelia, the fungal equivalent) deeply embedded in reality. *Cordyceps* and *Ophiocordyceps* are both genera of parasitic fungi. Both are in the order Hypocreales but belong to different families: Cordycipitaceae and Ophiocordycipitaceae, respectively. Some fungi in the *Cordyceps* genus, such as *Cordyceps sinensis*, have been used for centuries in traditional Chinese medicine—the fruiting body of the fungus, which is dried and ground into powder for human consumption, grows from the corpses of infected moth larvae that develop underground, and the fungus pokes through the soil like a tiny finger. *C. sinensis*, also known as Dong Chong Xia Cao ("winter worm, summer grass" in Chinese), is taken as an anti-inflammatory and has been used to treat a range of conditions: from respiratory illnesses and renal disease to low libido and fatigue.[1]

The genus *Ophiocordyceps*, on the other hand, is known for manipulating arthropod behavior. *Ophiocordyceps unilateralis* is particularly notorious and is commonly called the zombie-ant fungus. Druckmann, who began developing *The Last of Us* in the early 2000s while still an undergrad at Carnegie Mellon University in Pittsburgh, found inspiration for the game's fungal bosses after seeing *O. unilateralis* in action in an episode of the BBC nature documentary series *Planet Earth*. (The program, which first aired in 2006, refers to the fungus as "Cordyceps," which is how it was commonly identified at the time.)[2]

In the episode, a tropical bullet ant first shows signs that something is amiss by excessively grooming itself and then climbing unsteadily upward on a plant stem, where it eventually stops and bites down.

"Like something out of science fiction, the fruiting body of the Cordyceps erupts from the ant's head," says narrator Sir David Attenborough. His voice, beloved by fans of nature documentaries, is soothing; the footage is not. In a time-lapse sequence spanning weeks, a fungal stalk grows from behind the head of the deceased ant. When the stalk releases its spores," Attenborough warns, "any ant in the vicinity will be in serious risk of death."

For Druckmann, the sight of that zombie ant (see figure 2) inspired a chilling question about the behavior-manipulating fungus: "What if it jumped to humans?" For a person to be puppeteered as the bullet ant was—their mind destroyed and fungus in control of their body—would be a "fate worse than death," Druckmann told NPR in 2013.[3]

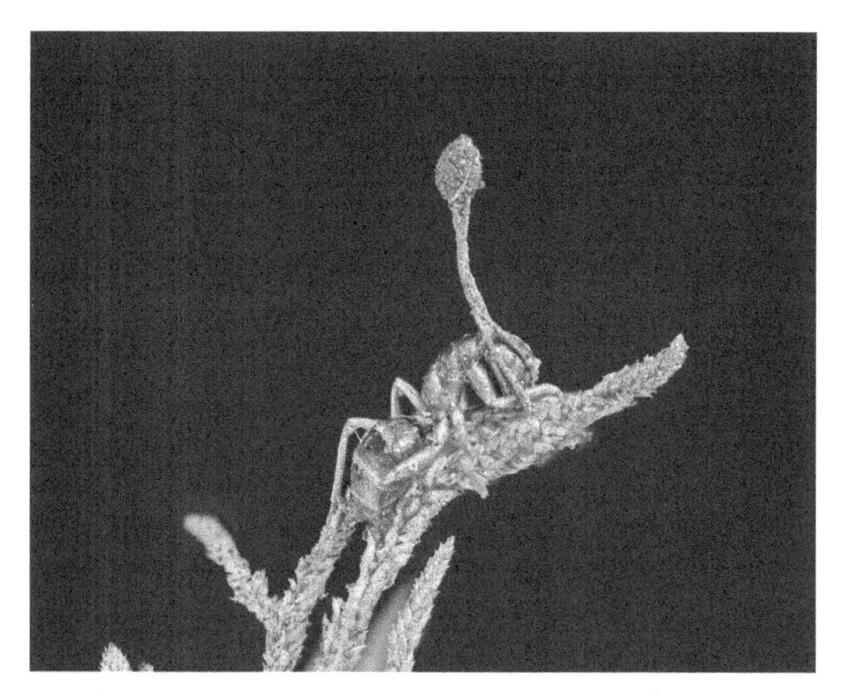

Figure 2 The zombie-ant fungus *Ophiocordyceps unilateralis s.l.* and one of its victims. *Courtesy of João Araújo*

To infuse the game with the horrific reality of *Ophiocordyceps* infection, the game developers consulted with David Hughes, then an assistant professor of entomology and biology at The Pennsylvania State University and a member of the university's Center for Infectious Disease Dynamics. Hughes already had over a decade of research under his belt investigating the relationships between parasites and their insect hosts, particularly when behavior manipulation was involved. He first worked with strepsipterans, but as fascinating as those twisted-wing parasites are, Hughes says, the group was at the time woefully understudied. Very little was known about their biology and genetics, let alone how they manipulated their wasp hosts.

Fungi were a different story. By comparison, far more was already known about their genomes and the chemicals that they produced. So when it came to unraveling how zombification happens, "the fungal toolbox was so much better to be able to get into mechanisms," Hughes says. And around 2006, while at the University of Copenhagen, Hughes shifted his focus toward *Ophiocordyceps* and rainforest ants.

Field observations and lab microscopy of infected ants enabled Hughes and others to map the diversity and reach of *Ophiocordyceps* in ants and to construct a directory of behavioral and physical changes in an ant while under the fungus's control, during the climbing and leaf-biting stages of infection. One such discovery, acquired through electron microscopy, 3D visualization, and analysis of images using deep learning algorithms, showed that the fungus extended throughout its host's body, forming networks that wrapped around the ant's muscles. But while the ants' heads were stuffed with fungal cells, *Ophiocordyceps* didn't extend its tendrils into the insects' brain tissue, hinting that other pathways were involved in controlling ant behavior.[4]

Hughes's research now focuses on global food security, and he has set aside his work on zombie ants, though there are dozens of other researchers worldwide exploring how pathogenic fungus zombifies insects. But when Hughes began his work with *Ophiocordyceps* in the early 2000s, "we knew basically nothing about it," he says.

The discovery of the zombie-ant fungus is often attributed to the nineteenth-century English naturalist Alfred Russel Wallace, a contemporary of Charles Darwin (independently of Darwin, Wallace also developed a theory of evolution by way of natural selection). Wallace collected ants infected by *Ophiocordyceps* fungus in 1859 while visiting the Indonesian island of Sulawesi (then known as Celebes), a place that he described as "the most remarkable and interesting [island] in the whole region, or perhaps on the globe." There, between 1854 and 1862, Wallace observed and gathered specimens representing thousands of species—many of which were new to science—publishing his notes in 1898 in the book *The Malay Archipelago*.[5]

But Wallace didn't write about the ants in his book. His name is mentioned alongside infected ant specimens in a record published in 1886 in *The Annals and Magazine of Natural History* describing specimens of *Cordyceps unilateralis* (as the behavior-manipulator *Ophiocordyceps unilateralis* was then known) in the British Museum collection. Two infected ant species, *Echinopoa melanarctos* and *Polyrhachis merops*, were collected "by Mr. A.R. Wallace at Tondano, a village in the Celebes," according to the record.[6] Presumably, Wallace found the ants clinging to rainforest leaves as *Ophiocordyceps*-infected ants do today—he may even have discovered many infected ant corpses clustered together, as is typical in locations marked by *Ophiocordyceps* contagion, Hughes says. But whether or not Wallace had any thoughts on that subject is anyone's guess.

Wallace and many of the scientists who followed after him and collected and studied infected ants didn't suspect that these afflicted insects were no longer ants—not exactly, according to Hughes. Rather, what these biologists were looking at was fungus walking around in an ant suit.

"It's not an ant," Hughes says. "It's an extended phenotype of the fungus."

One feature that makes zombie ants stand out to biologists is the dramatic neck-sprouting appearance of *Ophiocordyceps*—and that only appears once the ant is already a corpse. After Wallace, most of the early twentieth-century records of zombie ant parasitism focused on the fungus rather than on infected ant behaviors. But in recent decades, researchers began looking more closely at what happens before the ant dies, as it gradually becomes less of an ant and more an extension of the fungus.

Ophiocordyceps gains entry to its host as a single spore that sticks to the ant's exoskeleton, punches through the cuticle, and then pumps out new fungal cells that consume the ant from the inside. As the fungus grows, its structures surround the ant's muscle tissue; by the time infection reaches its peak, the ant's body mass may be up to 40 percent fungus.

About one to two weeks after a spore successfully infiltrates an ant's defenses, the fungus takes control, and the normally social ant deserts its foraging sisters. In an infected ground-dwelling ant, *Ophiocordyceps* guides its host away from its home; the ant wanders until finally it climbs to an elevated perch on a leaf or twig about 10 inches (25 centimeters) above the colony's foraging trails. A combination of chemicals and fungal structures that penetrate the ant's jaw muscles compel the insect to bite down hard. Once a zombie ant locks its jaws onto a leaf vein, *Ophiocordyceps* sets to work destroying the muscle fibers in its host's mandibles. With the ant now physically incapable of unlatching itself, the fungus is free to focus on the important work of reproduction.

The ant dies fixed to this elevated spot, but *Ophiocordyceps* isn't done yet. Fungal tendrils creep from seams in the ant's body. From the base of the dead ant's head, a slender stalk slowly, over two to three weeks, pushes through the ant's exoskeleton, lengthening upward until it sways over the corpse of its ant host—

which even in death still clings tightly to its perch with its jaws. A dark, egg-shaped, spore-producing lump called a stroma swells on this neck-stalk, then explosively releases a shower of spores. The spores drift down to the forest floor where they are picked up by foraging ants; these unsuspecting insects will soon become the next zombie victims.

In tropical ant species that are arboreal and nest high in the trees, such as *Camponotus leonardi*, a large species of carpenter ant, *Ophiocordyceps* follows a slightly different strategy. *C. leonardi* ants in a forest in Thailand, parasitized by *Ophiocordyceps unilateralis*, don't climb higher in the treetops. Rather, they descend from their nests in the hot, dry canopy and seek out vegetation closer to the ground. Under the influence of the fungus, an infected ant lurches down the tree trunk toward the humid understory, its journey frequently interrupted by convulsions that send it tumbling farther down the tree. When it finds a leaf at the appropriate height above the ground, it sinks its mandibles deep into leaf tissue around a vein, dying in place. Biting usually happens around noon, when the sun is at its peak overhead, Hughes and other researchers reported in *BMC Ecology*.[7]

It's unknown how long it takes the ants to die after initiating the death grip, since twitching in its limbs and body could be caused by growth of fungal masses inside the ant's corpse. "But it did appear that ants could remain alive for as long as six hours after biting," Hughes and his coauthors wrote. Microscopy of the dead ants' heads revealed that they were stuffed with fungal cells that surrounded the brain but did not penetrate it, and that their jaw muscles had atrophied—likely eroded by the fungus—keeping their bodies locked in place.

A similar zombie fate awaits the canopy-dwelling trap-jaw ant species *Daceton armigerum* when they're parasitized by *Ophiocordyceps daceti*. The ants live in rainforests across northern South America, and once they're infected, they too descend from the forest canopies and seek out leaves to bite in the humid patches of vegetation near the forest floor. From there, fungal spores will be readily transmitted when healthy ants descend from their canopy nests to forage among the nutritious plants and mosses.

"The fact that diseased ants are found in the litter and understory layers indicates a dramatic behavioural change following infection," Hughes and another team of scientists wrote in *Studies in Mycology*.[8]

Whether zombified ants stumble upward or stagger downward toward their mandible-clenching, leaf-gripping finale, the behavior contradicts their normal routine. And the fungus is the one in control.

*

You might not think you have much in common with a zombifying fungus, but fungi are more closely related to you than you may expect (on an evolutionary timescale, that is). Molecular analysis in the late 1990s showed that on the tree of life, the animal lineage split from plants about 1.547 billion years ago, while animals and fungi diverged around 1.538 billion years ago—approximately 9 million years later—making fungi closer kin to animals than to plants.[9] However, eighteenth-century scientists initially grouped fungi together with plants, based on guiding principles established by Swedish biologist and father of modern taxonomy, Carl Linnaeus. According to Linnaeus, all life on Earth could be lumped into two kingdoms: plants and animals. Shaping these categories was a pithy saying that has long been attributed to Linnaeus: "Minerals grow, plants grow and live, animals grow, live, and feel."

Mushrooms were the first fungi to be observed and classified; their rigid cell walls and the fact that they seemed rooted in place and didn't move around indicated to scientists of the day that fungi were plantlike. But fungi differ from plants in several key ways. Strengthening their cell walls is a material called chitin, which is found in animals but not in plants (plant cell walls are toughened by cellulose). Nor do fungi have chlorophyll as plants do, which means that fungi can't feed themselves by producing carbohydrates through photosynthesis. Instead, fungi absorb nutrients from organic matter. These and other features eventually led scientists in the late 1960s to reevaluate the group, and to finally move fungi into their own kingdom.

Over centuries of study, scientists learned that mushrooms were just the tip of the . . . well, of the fungi. Mushrooms belong to a fungus group that produces spores in visible fruiting bodies—reproductive structures sprouting from a network of threadlike hyphae, which absorb nutrients and aid in decomposition. Tendrils of hyphae weave together to form the mycelium, which is usually deeply buried in the soil or organism that feeds the fungus. Not all fungi produce these fruiting structures; yeasts, molds, and mildews don't, but they're fungi, too. To date, scientists have identified roughly 150,000 species of fungi (though by some estimates there are millions of species waiting to be found and described[10]), and the group has evolved numerous adaptations to survive in terrestrial niches almost everywhere on the planet, feeding on organic materials that other microbes can't easily process or can't break down as quickly.

Fungi can be unicellular—think of the yeasts that are commonly used for baking or brewing beer—but most are multicellular. They typically reproduce by dispersing spores, and their reproduction may be sexual, asexual, or a combination of both. A number of fungi produce a range of toxic compounds known as mycotoxins, many of which can be lethal to animals and plants. Some fungal compounds are used in medicines; the mold *Penicillium rubens* is the source of the antibiotic penicillin, and the fungus *Aspergillus terreus* produces lovastatin, which lowers cholesterol and reduces the risk of cardiovascular disease. In many cases, fungi are microscopic. But in northeastern Oregon's Blue Mountains, there's a single fungal entity that has existed for millennia, extending its flat, stringlike structures called rhizomorphs across thousands of acres of forest floor. Discovered in 1998, this gargantuan specimen of the fungus *Armillaria ostoyae* is estimated to be at least 2,400 years old (and possibly as old as 8,650 years). With a "body" spanning nearly 4 square miles (10 square kilometers), it's by far the largest organism on Earth.

Fungal feeding is diverse, too. Some fungi are saprophytic, feeding on dead and decaying organisms—plants, animals, and other fungi. Others coexist with certain organisms mutualistically, when the relationship benefits both partners. One example of this is mycorrhizae: fungi that grow in a plant's root system and serve as an extension of the root. The fungus helps the plant reach more nutrients; in return, the plant nourishes the fungus. Predatory fungi in the *Arthrobotrys* genus are hunters, trapping and devouring tiny worms called nematodes. Other fungal species are parasites and pathogens, surviving by extracting resources from living animals or plant life, and often harming their hosts. There are more than 600 fungal species that parasitize vertebrates, with more than 200 species associated with people. Some, such as the yeast *Candida*, create mild infections, while others—*Pneumocystis jirovecii*, for example, which causes a severe type of fungal pneumonia—can lead to more serious illness. Approximately 8,000 species of fungi cause disease in plants (around 400 plant species in turn parasitize fungi, which seems only fair). Roughly 1,000 species of fungi are known to parasitize insects.

And then, there are the zombifiers.

Ophiocordyceps, which first appeared about 100 million years ago, was designated as a genus in 1931 by British mycologist Thomas Petch. It includes around 320 species; of those, 35 species are known to not only infect their insect hosts but also to manipulate their behavior. And there may be hundreds more zombifying species in this group that are yet to be discovered, says João Araújo, a mycologist whose research focuses on insect-associated fungi. Araújo, whose arms are inked with colorful tattoos that include the fungal zombifiers that he studies, is a mycology curator at the Natural History Museum of Denmark and an assistant professor at Copenhagen University. Previously, he was the assistant curator of mycology at the New York Botanical Garden in the Bronx; with approximately 600,000 specimens in its fungus herbarium, it's home to one of the largest mycology collections in the Western Hemisphere.

In his office and in the collections in New York, Araújo showed me some insect victims of deadly fungus—ants, wasps, scale insects, moths, and others—that he and other scientists had collected during their expeditions; some specimens, packed in cardboard boxes with handwritten cards noting the date and location where they were found, dated to the nineteenth century. Ants that had been killed by zombifying *Ophiocordyceps* fungus had the genus's trademark slender stalk (and sometimes more than one) extending from their bodies. Other insects, like a moth felled by *Akanthomyces* (a nonzombifying genus in the Cordycipitaceae family), were almost completely covered in coatings of fungus, the shapes of their bodies obscured under an eerie alien landscape of fungal peaks and valleys. I had to peer closely to see the details in some of the specimens—several of the fungus-infected victims were so tiny that they could fit on a fingertip, with plenty of room to spare.

Most of the known *Ophiocordyceps* manipulators live in the tropics, but their overall distribution is global. Over the past decade, dozens of new species have been identified—many by Araújo—in China, Thailand, Australia, Ghana, and Canada, as well as in Japan, the United States, Brazil, and other Amazonian locations.

"Every time I go to the forest, I find new species," Araújo says. "Every single time."

Not that finding them is easy. When most mycologists or hobbyists go mushroom hunting, the fungi they seek are usually big enough to be easily spotted (if you know where to look). Zombifying fungi like *Ophiocordyceps*—and their insect victims—are many orders of magnitude smaller. Once a likely location is identified, "you have to stalk the place," Araújo says. For him, this typically involves hunkering down in a small plot of dense and humid Amazonian rainforest for an hour or more, peering at thousands of leaves.

But *Ophiocordyceps* isn't restricted to remote jungle habitats. *Ophiocordyceps kimflemingiae*, the first *Ophiocordyceps* species found in the United States, was spotted by South Carolina naturalist Kim Fleming in her backyard. (Araújo was the lead author of a paper describing the fungus; the study also identified 14 other newfound *Ophiocordyceps* species that parasitized ants in Australia, Brazil, Colombia, and Japan.[11]) The novel North American species caught the attention of senior study author David Hughes after Fleming posted an image of an infected ant on Flickr. Fleming's backyard turned out to be a bit of a hot spot for the fungus; she has since captured thousands of images of local zombifiers, including the species that now bears her name. Araújo had a similar stroke of luck when he was living in Gainesville, Florida; in his backyard, he found a specimen of *Ophiocordyceps camponoti-floridani* growing out of the head of a dead carpenter ant (*Camponotus floridanus*) that had sunk its mandibles into a palm leaf.

Considering that within the last century fungi were still grouped together with plants, it's not surprising that recent decades have seen mycologists reevaluate the connections between fungus groups and revisit how fungi are organized, as new species are described and older species are reexamined through genetic analysis. "Cordyceps" was once a term that was broadly applied among fungi; it referred to about 400 species of fungi in the family Clavicipitaceae that infected insects, killed them, and then popped spore-producing stalks out of their hosts' dead bodies. However, many of those fungi not only looked very different from each other but also inhabited different ecological niches.

In the mid-2000s, scientists from the United States, Thailand, and South Korea launched a molecular study to understand the relationships between all the species that were described as *Cordyceps*, extracting DNA from cultures and specimens and looking at multiple genes across 162 taxonomic categories. They divided all fungi previously labeled as *Cordyceps* into three families—Ophiocordycipitaceae, Clavicipitaceae, and Cordycipitaceae—with each lineage radiating from a different shared ancestor. Behavior-manipulating fungi—mostly in the *Ophiocordyceps* genus and mostly affecting ants—all landed in Ophiocordycipitaceae. The so-called true *Cordyceps*, in which species parasitize and kill insects but don't change their behavior, were moved into Cordycipitaceae.

"The species we knew as *Cordyceps* were distributed in these three families, and new genera were erected to accommodate them," Araújo says. The study, published in 2007, "was a game changer for nomenclature."[12]

But while it's not scientifically accurate to refer to behavior-manipulating fungi as *Cordyceps*, that old connection persists. Colloquially, "Cordyceps" is still often used in reference to zombifying fungi (as in *The Last of Us*). Even for scientists, the long-standing association can be hard to shake. Araújo says that he describes his research area as "taxonomy, systematics, and the evolution of cordyceps fungi" (that's "cordyceps" in lowercase and without the italics) because the term is still commonly accepted as a general descriptor for insect-parasitizing, stalk-sprouting fungi—even the behavior-manipulating ones, he says.

Ophiocordyceps fungus is extremely particular about who it zombifies. In lab experiments, some *Ophiocordyceps* species are capable of infecting and killing multiple host species. But when it comes to behavior manipulation, "one ant, one *Ophiocordyceps* species" is the reigning hypothesis. The notion that each parasitized species has its own unique zombifier hints to scientists how many zombifying *Ophiocordyceps* species may exist, even if the fungi haven't yet been discovered, Araújo explains. Data collected thus far indicate that in the Brazilian Amazon, more than 50 percent of carpenter ant species—ants in the *Camponotus* genus—are infected with their own, unique species of *Ophiocordyceps* zombifier. There are about 1,200 species of carpenter ants worldwide; it's possible that about 50 percent of those species also have a zombifying fungus that evolved to target them alone. That means there could be 600 or more *Ophiocordyceps* zombifiers lurking in ant populations around the world.

In fact, the species now recognized as *Ophiocordyceps unilateralis* has long been suspected of harboring multiple species that are yet to be identified—the name sometimes appears in scientific literature followed by *sensu lato* (*s.l.*), which means "in the broad sense" in Latin, hinting at its undiscovered diversity. In 2011, Hughes and two entomologists from the Universidade Federal de Viçosa in Brazil—Harry C. Evans and Simon L. Elliot—proposed that *O. unilateralis* represented a species complex (a closely related group of species that are genetically very similar) that might include hundreds of species.[13] Araújo and Hughes's 2018 study naming the South Carolina species (coauthored with Evans and US Department of Agriculture entomologist Ryan Kepler) also described 14 more *Ophiocordyceps* species, suggesting that they were part of the *O. unilateralis* "core clade"—a group of species specializing on Camponotini, the tribe that includes carpenter ants.[14]

In 2020, scientists from Taiwan reported finding *O. unilateralis* in eight different species of ants in an evergreen forest. The researchers had collected 281 dead ants in central Taiwan that showed signs of fungal growth, which they identified as *O. unilateralis*; there were seven ant species in the *Polyrhachis* genus and one in *Camponotus*. All showed signs that *O. unilateralis* hadn't just killed them—it had zombified them first.[15]

"The eight ant species all expressed death grip behavior to secure their place, leaving characteristic dumbbell-shaped scars centered around leaf veins," the Taiwanese researchers wrote. "This lockjaw trait or death grip is essential in the fungus's lifecycle." The manipulation was also very specific in where it guided the ants. *O. unilateralis* made the insects perform their death grip on plants only after reaching the desired height above the forest floor. There, conditions were more humid than in the forest canopy where the ants typically live, "to ensure an optimum microclimatic site for the next steps of fungal development and spore release."

In addition to the death grip they held on their leaves, many of the dead ants in the Taiwan forest were trussed to their perches by fungal mats sprouting from their legs, mandibles, and abdomens. However, the fungus hadn't invaded the leaves' tissues, suggesting that *Ophiocordyceps* wasn't feeding off the plants. Rather, it was getting all the nutrition that it needed from the corpses of its insect hosts.

Scientists are still piecing together how the fungus knows when it's found the singular species that it can zombify, but Araújo suggests that host recognition likely originates in chemical cues picked up by spores that settle on the ants and infiltrate their bodies.

"In order to trigger the spore germination, it has to be in contact with the cuticle of the right host," he says. "Once it gets to that situation, the spore will attach and then will germinate and penetrate." After a spore breaks through the ant's exoskeleton, the spore tip starts to bud, forming fungal cells that are released into the ant's body cavity.

Overall, there is still much to be learned about host specificity in *Ophiocordyceps*, and scientists have only begun to scratch the surface of the group's fungi-host relationship, Araújo, Hughes, and their coauthors wrote in 2018. But a key part of unlocking those secrets will depend on learning as much as possible about how widespread *Ophiocordyceps* really is.

"By unravelling the true diversity of this group, more intriguing and complex questions will come to the light," they wrote.

*

Chief among the questions scientists are asking is how exactly the fungus alters its hosts' behaviors. The molecular underpinnings of what turns an ant into a zombie have intrigued microbial ecologist Charissa de Bekker for over a decade. De Bekker, an assistant professor at Utrecht University in the Netherlands, was completing her PhD thesis in fungal genetics at Utrecht when she happened to see the *Planet Earth* zombie ant episode, which sparked an interest in *Ophiocordyceps* and brought de Bekker to Hughes's lab at Penn State, as a postdoc. She later led a lab of her own at the University of Central Florida, working to uncover the fungus's behavior-manipulating secrets. By piecing together how *Ophiocordyceps* changes ant behavior on a molecular level, de Bekker told me, she hopes to find answers about the molecules that regulate normal animal behavior, too.

In 2014, de Bekker and other researchers from Penn State (Hughes included) sampled a newfound species of *Ophiocordyceps unilateralis s.l.* found in two species of carpenter ants in South Carolina (*Camponotus castaneus* and *Camponotus americanus*) and cultured the fungus in the lab. That accomplishment—isolating the fungus from a *C. castaneus* corpse, growing it under controlled conditions, and replicating the infection in ants by injecting them with fungal cells—was a significant breakthrough, de Bekker says.

"We even threw a baby shower for the fungus when I finally isolated it," she recalls. "We had cake because I finally was able to grow some fungus in the plate."

The scientists then experimentally infected worker ants in the host species as well as two other ant species that share habitats with *C. castaneus* and *C. americanus* but had never been found hosting the fungus in the wild (the researchers confirmed this by observing the unafflicted species—*Camponotus pennsylvanicus* and *Formica dolosa*—for four years, for a total of about 1,750 hours).

In the lab, *O. unilateralis s.l.* killed all the ants that it infected, regardless of species. However, the fungus induced end-of-life height-seeking and twig-biting only in the two species that it infected in nature, with the changes appearing about two to three weeks post-infection. Chemical comparisons of tissues in the infected ants—those that *O. unilateralis s.l.* manipulated and those it didn't—gave the researchers candidate metabolic compounds, or metabolites, that might play a role in manipulating ant brains. Amid thousands of chemicals that the scientists recorded and reviewed, two known brain-changers stood out in the favored host species: guanidinobutyric acid (GBA) and sphingosine. Spikes in these neuromodulators were previously associated with behavior disorders in mammals. But as these compounds are already present in animal cells, the study authors were unsure if the fungus was itself producing the compounds or was inducing the ants to do so by flooding its brain with other types of molecules, which were as yet unidentified.[16]

Another study the following year, also led by de Bekker and with Hughes as senior author, targeted gene expression in *O. unilateralis s.l.* as it infected *C. castaneus* ants; the research team, which included scientists from universities in the United States, Germany, and the Netherlands, checked more than 7,000 genes, searching for the ones that were coding instructions for making proteins while manipulation was underway.[17] Of the fungus's genes that were active during host climbing and biting, 80 percent of those genes were absent from the genomes of nonmanipulating fungi. Proteins that the fungus produced changed host levels of serotonin, which is linked to foraging behavior in ants, and dopamine, which is associated with biting. Other proteins associated with the fungus were thought to disrupt chemoreception, blocking an infected ant from following established foraging trails.

"We also don't see them really communicating with other individuals anymore," de Bekker says. "That also, for ants, happens mostly through smell."

The researchers also detected fluctuations in an enzyme called tyrosine 3-monooxygenase, previously known to increase activity in insects and to regulate circadian rhythms—body processes and behaviors that respond to the day-night cycle. Dosing an ant with this enzyme may help the fungus control the timing of the host's final doomed bite, which in many infected species takes place at midday. The enzyme may also contribute to the ant's highly active wandering stage, driving it away from the nest before it bites down on a plant and dies.

Raquel Loreto, a researcher with Penn State's Department of Entomology and Center for Infectious Disease Dynamics, also investigated zombie ants during their plant-biting phase; together with Hughes, she created a metabolic profile of their brains. While the brains of ants infected by *Ophiocordyceps kimflemingiae* were free of fungal structures, they were "notably different" from the brains of healthy insects—in their energy use, chemical makeup and overall health, Loreto and Hughes found. For example, manipulated ants' brains had seven times less of the neurotransmitter adenosine, an important regulator of normal function in the central nervous system, and about ten times more hypoxanthine, which has been linked to cognitive, behavioral, and neurological dysfunction.[18]

Not all of the fungal intervention in the ant's brain was destructive. Elevated levels of a fungal compound called ergothioneine, which guards cells against damage, suggested that the fungus was also protecting the brain of its zombie host—that is, until the host died. After that, its brain was just another entrée on *Ophiocordyceps*'s dinner menu.

Each metabolic and genomic discovery sheds light on how *Ophiocordyceps* zombifies ants and provides scientists with tools for understanding the mechanisms of other zombifiers, according to a 2023 analysis of *Ophiocordyceps* compounds that affect ant stability and walking.

"Integrative behavioural studies like this one," the authors wrote in *Animal Behavior*, "bring us one step closer to characterizing the molecular mechanisms that drive not only zombie-making fungi, but other behaviour-manipulating parasites across the Tree of Life."[19]

*

How long ago did fungi evolve tools for zombification? Evidence of any type of parasitic association is exceptionally rare in the fossil record. The oldest direct evidence of fungi parasitizing insects dates to the Early Cretaceous and was discovered in 2008, preserved in a tiny lump of amber from Myanmar (formerly Burma) measuring 0.3 inches (8 millimeters) long and 0.2 inches (6 millimeters) wide. In the fossil, fungal reproductive structures called synnemata extend from the body of a scale insect, so the scientists named the fungus *Paleoophiocordyceps coccophagus* (the species name means "scale-eater" in Greek). Their analysis suggested that the major lineages of fungi in the Hypocreales order originated in the Late Jurassic and then diversified in the Cretaceous, and that the relationship between the Ophiocordycipitaceae family and its insect victims is at least 122 million years old.[20]

A somewhat younger fossil example is a piece of Baltic amber that's around 50 million years old. Found in Russia's Samland Peninsula, the amber tomb preserves a fungus-infected ant with a stalk-like structure emerging from behind its head. Scientists named the fungus *Allocordyceps baltica*, proposing that it could be an ancestor to the *Ophiocordyceps* genus.[21]

As for fossilized signs of parasitic behavior manipulation, such evidence is almost nonexistent—with one notable exception: dozens of dumbbell-shaped scars in a fossilized leaf dating to about 48 million years, identified in 2010 by Hughes, Torsten Wappler (head of the natural history department at the Hessisches Landesmuseum Darmstadt in Germany), and Conrad Labandeira (curator of fossil arthropods at the Smithsonian Natural History Museum in Washington, DC). These marks hinted at fungal manipulation of insects in a manner similar to the biting behavior seen in *Ophiocordyceps*-infected carpenter ants at the end of their lives. Along 11 secondary veins in the fossil leaf, found at the Messel Pit site in Germany, were 29 marks resembling those made in modern-day plant tissues by the deathgrip of zombie ants.

Hughes, Wappler, and Labandeira compared the ancient marks to bites made by modern *Camponotus leonardi* worker ants killed by *Ophiocordyceps unilateralis*; the dead ants were found clinging to the undersides of leaves in the Khao Chong Peninsular Botanical Garden, in southern Thailand. The bite marks that they made were photographed and then compared under high magnification to the marks that were made 48 million years ago. In their shape and locations on the leaves, the damage from recent and ancient scars was a close match.

"The dating of this unique parasitic association minimally to the mid Paleogene [66 million to 23 million years ago], indicates a deep-time origin for this phenomenon," the scientists reported in *Biology Letters*.[22] By that time in the Paleogene, ant lineages had already diversified and multiplied significantly across ecosystems worldwide. Ant zombification by fungi—a highly specialized interaction—is therefore "relatively ancient" in origin, the study authors wrote.

While biting plants is a key end-of-life behavior for *Ophiocordyceps*-infected ants, what they bite can vary depending on where in the world they live. Infected ants in temperate habitats, such as those in South Carolina, zombified by *Ophiocordyceps kimflemingiae*, are found gripping twigs rather than leaves, which makes sense from the fungus's point of view.

"We found out that these species required a whole year to complete their life cycle," Araújo says. In habitats where trees shed their leaves in the fall, a leaf-biting ant corpse would end up on the ground and buried under leaf litter, robbing the fungus of its chance to poof out spores onto new victims. A twig-biting ant, on the other hand, remains securely in place even as leaves drop—such ants are often found "hugging" the twigs with their legs, as well as grasping with their jaws (see figure 3). The ants' corpses freeze and then defrost as the seasons change, until the fungus is ready to reproduce.

Ants may be the best-known of all *Ophiocordyceps*'s insect targets, but they aren't its only victims. *Ophiocordyceps* hosts are spread across ten insect orders and are predominantly found in four: *Coleoptera* (beetles), *Hemiptera* (true bugs), *Hymenoptera* (ants, bees, wasps, and sawflies), and *Lepidoptera* (moths and butterflies). The fungus may be even more widespread than that—in 2021, scientists found the first evidence that *Ophiocordyceps* species can infect insects in *Blattodea*, the order that includes termites and cockroaches.[23]

However, when it comes to the sheer number of individual insects that are available for zombification, ants are unparalleled. Some species have tens of millions of insects in a single colony. In tropical forests, ants are so abundant that they often represent 80 percent of all arthropods; by some estimates, if all of a rainforest biome's animal life were gathered together and weighed, nearly 50 percent of the collective biomass would be ants. On a global scale, a recent study suggested that there are a staggering 20 quadrillion ants on Earth—sweep them up into a giant ant ball and you'd end up with a mass that would weigh approximately 12 megatons. To put that in perspective, that's more than the combined biomass of all mammals and birds in the

Figure 3 In temperate ecosystems, *Ophiocordyceps*-infected ants climb and bite down on twigs. *Courtesy of João Araújo*

wild (and about 20 percent of the biomass of all humans on the planet).[24] Ants' abundance—especially in tropical forests—is part of their success story as an insect group; and it's also why they're so commonly infected by *Ophiocordyceps*.

In places where the fungus is successful—usually close to established ant colonies—the undergrowth on the forest floor dangles scores of tiny, fungus-sprouting ant corpses on the undersides of leaves, their jaws clamped tight around leaf veins. This is extremely helpful to scientists who study the tiny infected ants because "if you find one, there's lots more in that specific area," de Bekker says. "But then if you venture out of that area, you might not find anything." Death-stinations that are favored by *Ophiocordyceps* are usually selected because the temperature and humidity there are ideal for fungal growth inside ant cadavers; the neck-sprouting stalks need time and moisture in order to develop to the point where they can expel their spores. Such locations are typically near workers' well-trodden paths, so that the eventual spore-showers will be certain to intercept foraging ants.

By sending its ant host stumbling away from the nest on a death march, the fungus also ensures that a soon-to-be corpse won't end up on the colony's trash heap, tossed there by a fastidious nestmate during her maintenance shift. Corpse-removing behavior, known as necrophoresis, is important for nest hygiene and is common in social insects such as ants, termites, and honey bees. Ants are careful about monitoring contagion in the colony, and dead ants are bundled up and carried to a spot where pathogens won't spread to healthy individuals. Some species carry their dead to refuse piles outside of the colony, while other species bury their dead or set aside a chamber in the nest for corpse storage.

Ant colonies may designate their own cemeteries, but these don't suit the purposes of *Ophiocordyceps*—the fungus wouldn't have much chance of infecting more ants when its host's body is isolated from the colony or buried. And so *Ophiocordyceps* guides its zombies to a very different type of dead zone out in the open. At the fungus-selected places where infected ants dangle, the corpses sometimes accumulate in numbers so great that scientists refer to the sites as "graveyards," and cadaver density can exceed 25 ants per square meter.[25] Finding such graveyards is painstaking work; in 2009, researchers from Denmark, Thailand, and the United States (Hughes was senior author of the study) mapped aggregations of ant cadavers in southern Thailand by turning over every single leaf in a rainforest plot measuring more than 14,600 square feet (1,360 square meters). The patience of their efforts paid off. In total, they counted 2,243 dead ants, each anchored in place by its clenched jaws and sporting a telltale stalk behind its head, a monument to the effectiveness of *Ophiocordyceps* manipulation in the name of reproduction.[26]

The fungus may be guiding its hosts to such spots by affecting how they perceive light. In 2019, de Bekker collaborated with biologists in Brazil who tested that hypothesis by experimentally changing light exposure in rainforest locations where *Ophiocordyceps camponoti-atricipis* had established graveyards for the carpenter ant *Camponotus atriceps*. The scientists set up shading screens over ten sections of ant graveyards in the Brazilian Amazon's Ducke Forest Reserve, leaving half of each section in shadow. They counted existing ants and measured their positions, and then waited for new graveyard occupants to show up.

Newcomer zombie ants visited six of the ten plots, and were overwhelmingly drawn to the sunnier side of the graveyards. "The presence of dead infected *C. atriceps* was strongly influenced by experimental light reduction," the researchers reported in *Behavioral Ecology*. "Shaded areas harbored fewer recently infected ants compared to naturally illuminated areas." In the ants that died in the shade, fewer produced *Ophiocordyceps* fruiting bodies. Those ants also climbed higher than zombie ants in sunnier spots, likely compelled by their fungal puppeteer to work just a little harder to find a more suitable light level before expiring.

"This suggests that the high density of dead ants in graveyards might be attributed to the incident light in these areas," the study authors wrote.[27] And the presence of light likely accompanies a certain range of temperature and humidity that the fungus needs for reproduction, de Bekker says.

The arrangement of all those tiny ant bodies in a zombie ant graveyard is unquestionably the stuff of nightmares, but there may be a silver lining to the fungal ant-pocalypse—for organisms other than ants, that is. Fungi of all types contribute to the health of forest ecosystems, fueling decomposition and dispensing vital nutrients to plants and soil. By zombifying ants, fungi may also be providing a helpful service; without *Ophiocordyceps* keeping some tropical ant populations in check, those species might otherwise overwhelm their habitats.

At the same time, *Ophiocordyceps* doesn't seem to wipe out its host ant populations. In this very important way, says de Bekker, the real world deviates from zombie narratives in which humanity is eradicated—or nearly so—by a zombifying pathogen. In fact, when scientists from the United States and Brazil conducted long-term observations of Brazilian *Camponotus rufipes* ants, which are commonly infected by *Ophiocordyceps camponoti-rufipedis*, they found that 100 percent of the ants' colonies in the study area harbored individuals that were infected by the fungus. But those infection rates were low; over 20 months, the average death rate was only about 15 ants per colony per month. No colony was snuffed out, but neither did any manage to purge itself of the contagion, suggesting that fungal presence was consistent but not a threat to the colony as a whole.[28] Older individuals—the foragers—are the ones who get exposed to the fungus. And while the sight of ant graveyards may look dramatic to human eyes, they typically represent less than 2 percent of a colony, Hughes says.

"It's kind of like a chronic cold, in the background," de Bekker adds. "It's keeping numbers in check, but it's never making an entire colony collapse."

*

If only the humans of *The Last of Us* were so lucky.

The HBO series wasted no time setting the stage for how humanity might be wiped out by a zombifying fungus. In the first episode, the first scene (set in 1968) features epidemiologist Dr. Neuman (John Hannah) warning a television talk show host that fungi are the likeliest source of the next global pandemic. While most fungi are ill-suited to survive the warmth of a human body, Neuman explains, climate change could drive fungi to adapt to withstand higher temperatures.

"So, if that happens . . . ," the discomfited host prompts.

"We lose," Neuman responds.

Later in the episode (fast-forwarding to 2003), the predicted pandemic is underway, causing chaos and societal collapse. The next episode, which culminated in Tess's horrific "kiss," offered more details of the pandemic's origin, tracing the human-infecting fungus to a mutated "Cordyceps" species that infected a worker in a flour and grain factory in Jakarta, and then spread globally.

Species adaptations by way of mutations and natural selection have been hallmarks of evolution ever since the appearance of the first microorganisms, about 3.7 billion years ago. However, the evolutionary shift in fungi that leads to human zombification in *The Last of Us* is unrealistic for a fungus that's best-known for zombifying ants.

Travel back into *Ophiocordyceps*'s ancestral past, before insects were on the menu at all, and you'll find a fungal ancestor that consumed plants. Crawling over those plants were all kinds of herbivorous arthropods; it wouldn't have been much of an evolutionary stretch for a plant-eating fungus to expand its culinary horizons and start sampling insects that were full of digested plant matter, Araújo says. In this way, a hypothetical lineage of fungi that started out as plant-dependent, at some point dabbled in and then switched to be entomopathogenic, or insect-infecting. They developed a taste for arthropod prey and never looked back.

"Once they conquered the insect body, they found a very nice niche to establish themselves in," says Araújo.

After getting comfortable parasitizing one insect lineage, it would have been comparatively easy for the fungi to continue adapting so as to parasitize other insects, too. As for when fungi started manipulating their hosts, Araújo and Hughes in 2019 pinpointed the origins of zombification for *Ophiocordyceps*. They first used genetic analysis of more than 600 species in the fungal order Hypocreales (which includes *Ophiocordyceps*) to establish context for behavior manipulation in the genus. They then created a family tree for *Ophiocordyceps* and identified a common ancestor to 129 species: a fungus that infected beetle larvae but didn't manipulate them.

It's possible, Araújo and Hughes wrote in *Current Biology*, that the nonzombifying fungus jumped from beetles to the ants that lived alongside them in soil or decaying wood. At first, the fungus probably attacked its ant hosts as it did beetles: by feeding on them and eventually killing them without altering their behavior. But unlike beetles, ants are social insects with behaviors that protect individuals—and the entire colony—against pathogens. Sick ants are typically killed and their cadavers destroyed or dumped away from the colony. The need to evade the "robust social immunity" of such hosts would create selective pressure for the fungus to actively subvert normal ant behavior. Making an infected ant leave its colony and secure itself to a leaf before dying would then enable the fungus to reach its spore-producing stage. By shifting from beetles to social insects, the fungus not only unlocked a new toolkit for host manipulation; it also prompted significant diversification in *Ophiocordyceps* that specialized in ants, Araújo and Hughes said.[29]

However, *Ophiocordyceps* may have diversified and specialized a little *too* well. In evolving behavior-manipulating mechanisms that are specific to just one host species, *Ophiocordyceps* zombifiers painted themselves into a corner (evolutionarily speaking) by limiting their opportunities to reproduce across a broader range of hosts.

"In order to apparently be able to change behavior in such a specific way, you have to have some sort of very tight coevolution with that host," de Bekker explains. "And it comes at a cost, right? The cost of not being able to be a generalist anymore."

Because *Ophiocordyceps* has become so highly specialized over millions of years of evolution, its reproductive reach at present doesn't even extend across all insects, let alone pose a threat to animals that are only very distantly linked to arthropods on the tree of life. While host-hopping is relatively common among fungi in the Hypocreales order—and has been since the order originated between 158 million and 232 million years ago—making the jump from manipulating arthropods to manipulating warm-blooded animals that are much larger and have more complex cells and nervous systems, would be an extreme evolutionary leap indeed. Fungi can be dangerous to people who are immunocompromised, but in general the group doesn't fare so well in our warm mammalian bodies; mammals may even have evolved high body temperatures as protection against fungal infection.[30] We inhale millions of invisible fungal spores every day, usually with no harmful impacts, and of the millions of fungus species on Earth, only a few hundred cause disease in people. Humans are most vulnerable to pathogens that evolved to infect our mammal relatives; diseases that sicken mammals (especially primates) can evade human immune responses more effectively than a fungus that typically attacks insects. And unlike viruses, which can easily be transmitted in aerosols, fungal infections typically don't spread person-to-person.

That said, there are fungi that have already adapted to live in humans as part of our microbiome—the community of mostly benign microbes that inhabit our bodies. Hundreds of species of fungi have been identified that are associated with the human digestive system. Some, such as the yeast *Candida albicans*, are opportunistic pathogens, which means that they're harmless most of the time but can multiply to pathogenic levels given the right circumstances. However—at least for the foreseeable future—humans are unlikely to fall victim to fungal zombification by way of a mutation in *Ophiocordyceps*.

Unfortunately for insects, the fungus kingdom holds myriad zombifying dangers—and not just from *Ophiocordyceps*. Many other parasitic manipulators lurk in the shadows, biding their time and waiting for an opportunity to seize control of an unsuspecting arthropod host and make it their unwilling puppet. In most cases, a zombie has no clue that it's been zombified until its behavior is changing, and by then—as Tess from *The Last of Us* would probably tell you—it's already much too late.

3

FUNGUS AMONG US, PART II

Zombie Flies, Beetles, and Millipedes

"They just want to live. Everyone wants that."
—MELANIE (SENNIA NANUA), *The Girl with All the Gifts*

At first glance, the insect clinging to a gently swaying blade of grass in a moonlit field resembles a resting bee, its body striped in yellow and black. With its legs and wings spread wide, it looks as though at any moment it might take off and fly back to the hive. But the insect isn't going anywhere, and closer examination reveals that it's no bee. It's a fly—a dead one—and the jaunty pale-yellow "stripes" are in fact spore-producing fungal masses that grew inside the fly and then burst through its body at seams in its abdomen. Before that, on the last evening of the fly's life, the fungus seized control of the insect's mind, compelling it at dusk to climb to an elevated perch. Over the hours that followed and shrouded by darkness, fungal spores erupted from the body and drifted down around it, forming a halo of death to ensnare the deceased insect's unsuspecting fly friends and neighbors.

The spread wings might mimic a pose of health and vigor, but they're just another example of how the fungus commandeered the body of its zombie host at the end of its life. Fanning the dying fly's wings wide moved them out of the way of the fungal structures sprouting from the insect's abdomen, providing the spores with more surface area for escape. And while the fungus isn't able to fully animate the dead and make a deceased fly host lurch toward a new victim, as do the walking dead in most zombie movies, this fungal zombifier has a few tricks that help it coax living flies into approaching an infected corpse. A dead fly's plump abdomen is a tempting sight to male flies searching for mates; they don't realize that their love interest's belly is bloated with spores and spring-loaded with death. Further confusing the issue are chemical signals released by the fungus, which mimic mating-associated scents and further entice male flies to approach close enough for the fungus to infect them and make new zombies.

Fungal zombifiers belong to a vast kingdom of organisms encompassing many species that reproduce by infecting—and in some cases, mind-controlling—a wide range of insects at every stage of their life cycles. We've already spent some quality time with the zombifying genus *Ophiocordyceps*, but others in the fungus kingdom are equally capable of gruesomely overpowering arthropods, transforming their bodies, and rewiring their behavior. As widespread and diverse as fungi are, the scope of this vast group is likely far greater than we know. Some estimates, drawing from associations of known fungal species with their hosts, suggest that globally there may be between 2.2 million and 3.8 million species of fungus. Other predictions are even more generous, placing global fungal diversity between 11.7 million and 13.2 million species. Based on those estimates, approximately 1.5 million species of fungus worldwide are thought to infect insects.

One fungal group that is particularly deadly to insects is the order Entomophthorales. It contains over 220 species, most of which are arthropod-parasitizing. The order's name is derived from a genus in the group—*Entomophthora*—which was named in 1856 from the Greek words *entomon* ("insect") and *phthora* ("destroyer"), by the German physician and botanist Johann

Baptist Georg Wolfgang Fresenius. The aptly named insect destroyers are found around the world, and the *Entomophthora* genus contains 21 species.[1]

The first *Entomophthora* species to be identified was a fly destroyer: *Entomophthora muscae* (it was originally classified as *Empusa muscae* and renamed a year later, as the genus *Empusa* was already in use as a classification for orchids). *E. muscae* debuted in scientific literature in the mid-nineteenth century under the pen of Ferdinand Julius Cohn, a German naturalist. Cohn was a botanist and a pioneer of microbiology and is perhaps best remembered today for his groundbreaking work with bacteria—he proposed the first system for classifying bacteria based on their shape. But in mycology circles, he's known as the first scientist to describe the fly-destroying and fly-manipulating *E. muscae* fungus.

Cohn published a report of *E. muscae* parasitizing common house flies (*Musca domestica*) in 1855, while he was studying botany at the University of Breslau in the German state of Prussia (the institution is now the University of Wroclaw, in Poland). He didn't have to travel far to collect his infected insect subjects; Cohn found plenty of fungus-stuffed flies clinging to window curtains in his home. In fact, he wrote, the sight of such flies was not at all unusual to the average German homeowner. The flies' "strange way of death" was "known to every child," though naturalists had yet to record the phenomenon, Cohn reported (these and subsequent excerpts from his 1855 article are translated from German).[2]

Despite its common appearance in houseflies, *E. muscae* infection was still "one of the strangest and most interesting apparitions" that Cohn had ever seen. And he spared no detail in his published description.

During the earliest stages of infection it's nearly impossible to tell that anything is wrong with the fly, Cohn wrote. But as the fungus inside it continues to grow, the insect becomes increasingly sluggish and won't fly away when approached.

"About an hour before death all locomotion ceases; the animal sucks itself tight with its proboscis," Cohn continued. The fly's legs twitch and its abdomen swells, turning a whitish color. Eventually, "the movements of agony cease; the animal no longer reacts to external stimuli," he wrote. After a fly died, its abdomen continued to expand, until a "white substance" pushed its way through the exoskeleton in three wide bands, emitting a faint scattering of "dust" on the ground surrounding the insect's corpse. The bands of white striping the fly's body slowly grew in width and height, and the density of the powdery substance on the substrate around it steadily increased.

"Gradually the body dries up, the white rings disappear, the stretched body shrinks," Cohn wrote, "and the fly almost assumes its usual appearance." But lingering traces of what killed it remain as a layer of powder coating the dead fly's wings and legs.

Cohn also created visual records of the fungus and its victim, in pencil illustrations drawn with a camera lucida. With this optical device, once commonly used by microscopy artists to illustrate very small objects and organisms, Cohn sketched the fungus at various developmental stages. One drawing of an infected fly showed the insect's abdomen ringed with pale circles that represented the fungus as it poked through the surface of the corpse.

Over the next century, scientists described dozens more *Entomophthora* species, many of which were found to be specific to a single host species or to a small group of close relatives. There's *Entomophthora culicis*, a scourge of the *Aedes aegypti* mosquito; the aphid-destroying *E. chromaphidis*; and *E. syrphi*, which affects hover flies (but not house flies), to name just a few. *E. muscae* was initially thought to affect an unusually broad range of fly hosts, but researchers have recently suggested that when

Cohn described *E. muscae* he mistakenly included observations of three other *Entomophthora* species that had similarly shaped spores: *E. pelliculosa*, *E. scatophagae*, and *E. syrphi*.[3]

Most *Entomophthora* fungi can manipulate insect behavior from inside hosts' bodies as part of reproduction, which culminates in asexually produced bell-shaped spores known as conidia. A hallmark of *Entomophthora* infection is a grand finale that forcibly expels spores from the zombified host in an explosive poof, a phenomenon known as a conidial shower.

"Broadly speaking, nearly all *Entomophthora* fungi follow a common survival strategy consisting of infecting, consuming, and then behaviorally manipulating their insect hosts," wrote Carolyn Elya, a research associate in organismic and evolutionary biology at Harvard University, and Henrik de Fine Licht, an associate professor at the University of Copenhagen in the mycology journal with arguably the best journal name of all time, *IMA Fungus*.[4]

Elya studies the processes that parasites such as *Entomophthora* use to change behavior, looking at the molecules, genes, and cells that make zombification possible. Her interest in *E. muscae* was kindled in 2015, while she was in graduate school at the University of California, Berkeley. The encounter was accidental, she told me; Elya found dead, infected insects amid decaying fruit that she was keeping in her backyard to attract wild fruit flies, for experiments unrelated to zombification.

"These dead flies had their wings at a 90-degree angle," she recalls. They also had a distinct banding around their bodies of a powdery substance. "I usually call it schmutz," Elya says. She suspected the schmutz (a Yiddish term for "dirt") was *Entomophthora* fungus, and she confirmed its identity with molecular tests.

Elya realized that having a backyard site that was an *Entomophthora* hot spot presented a unique opportunity. She knew that *Entomophthora* typically kills flies around sunset. By repeat-

Figure 4 After *E. muscae* kills its host, the fungus bursts through seams in the insect's exoskeleton. The fly's lifted wings help the fungus to disperse more freely. Image captured through a microscope (length of a fruit fly is 3 mm, on average). *Courtesy of Carolyn Elya*

edly dredging her backyard stash of decomposing fruit soup after sundown, Elya could collect insects that were freshly dead and still infectious. Back in the lab, fungus samples from those fruit flies could be used to infect other flies, enabling Elya to study the early stages of infection and learn which molecular tools the fungus kept in its zombifying toolbox (see figure 4).

"I've always just been really struck by the circadian element to the system," Elya says. "It causes these wonky behaviors. But then on top of that, all of this happens with this very orchestrated timing. I literally have set my clock to it for the past eight years."

Entomophthora affects insects across five orders: Diptera, Hemiptera, Plecoptera, Raphidioptera, and Thysanoptera. The fungus's relationship with these five insect groups likely originated in the Devonian period, around 419 million to 359 million years ago, as insects in these orders last shared a common ancestor around 400 million years ago.

Infection by *Entomophthora* fungus typically follows the same trajectory across the genus. From the swollen, spent corpse of a dead, infected insect, the fungus that killed it launches a fountain of adhesive conidia. Spores require direct contact with a healthy insect to progress to the next stage of reproduction. Once a spore touches down on an insect's cuticle—the outermost layer of the exoskeleton—it extrudes filaments that work their way inside the new host's body and swiftly begin to multiply in the blood-like fluid called hemolymph. At this stage in the fungus's life, it pumps out cells that are protoplasts—tiny blobs of protoplasm without a cellular wall. Cell walls produce antigens, so this adaptation is thought to help the sneaky fungus evade detection and destruction by the host's immune system. *E. muscae*'s fast growth in a still-living insect may also play a part in protecting the fungus from being ripped apart by an immune response.

Thread by thread, the fungus multiplies; as it does, it feeds off its new host. It dines on free-floating hemolymph nutrients and on fat tissues that won't immediately be missed by the insect, but it leaves vital organs and muscle tissue alone, as it needs the host to stay alive and mobile as long as possible. When this internal pantry is almost entirely depleted, the fungus prepares to take hold of its host's behavior. Protoplasts migrate into the insect's organs and start to develop walls. Zombification of the host is about to begin.

Some *Entomophthora* species produce thick-walled resting spores, which can weather exposure to the elements for months or years and then disperse later, with the arrival of a suitable host. For optimal dispersal of these resting spores, the fungus manipulates its insect zombies into heading toward the ground. Most *Entomophthora*, however, produce conidia. These spores ideally require elevation for dispersal. In these cases, the fungus compels its zombie host to climb to a high spot and hold on so tightly that even after death it won't easily be dislodged—at least, not before the fungus spews its spores into the surrounding area, launching them with structures called conidiophores in an explosive blast.

These host behaviors, in which directed movement takes place during the final stage of fungal infection, appear to serve the fungus alone by positioning the host in an ideal position for spore ejection, Elya and de Fine Licht wrote in *IMA Fungus*. This leaves little doubt that the fungus is directly manipulating its host for its own benefit. *Entomophthora* manipulation also follows very specific timetables; insects typically start their death marches as evening approaches so that the fungus can begin expelling its spores at night. This, too, benefits the fungus, as nighttime is when dew begins to accumulate, providing *E. muscae* with moist conditions that are preferable for germination.[5]

The *Entomophthora* group's flagship species, the body-busting, fly-destroying *E. muscae*, is now recognized as a parasite of more than 20 fly species across several families. Outbreaks of the fungus are most likely to occur in spring and fall, when conditions are cool and humid and there are abundant potential hosts. During such outbreaks, dead, fungus-filled flies can often be found clustered together at elevated spots, in places wherever the flies tend to naturally collect in groups. Zombie insects may gather on window frames inside homes. Outdoors, they cling to fences, barn walls, and other agricultural structures, and to tall garden plants, such as phlox, goldenrod, and thistles.[6] After a fly dies in an elevated location, *E. muscae*'s conidiophores push through seams in the cuticle and begin pumping out thousands of spores, in bursts that continue for the next 24 to 36 hours.[7]

When *E. muscae* spores land on a fly (and they can stick to any part of its body), chemical cues in the cuticle may help the fungus recognize if the species is one that the fungus can infect. Only an adult fly will do—*E. muscae* infections in fly eggs or larvae are unknown in the wild. Experimental attempts to introduce the fungus in any life stage other than an adult fly have been unsuccessful, Elya and de Fine Licht reported, though it's an open question as to why that might be the case.

"As with host specificity, the basis for life stage specificity is also unclear," they said. Perhaps *E. muscae* can only infect adults because its conidia are unable to penetrate the exterior of a larva or pupa, which differ compositionally from the cuticle of an adult fly. Another possibility is that some conidia are able to break through larval and pupal outer layers but then die because they can't find the nutrients that they need or because the chemical cues that trigger their growth when they infect adult flies are absent in the flies' earlier life stages.

If a spore wins the host lottery, within a few hours it germinates on the insect's body, producing thin filaments called germ tubes which use a combination of physical force and digestive enzymes to penetrate the fly's cuticle. The most frequent entry point is the abdomen—the body part that holds the most fat reserves—but scientists have found germ tubes poking through flies' legs, heads, and thoraxes, and even very occasionally penetrating veins in the wings.[8] As soon as the cuticle's tough shell is cracked, the germ tube burrows inside and begins producing hyphae, the filaments that make up the fungus's rootlike structure, the mycelium. Thread by thread, the hyphae extend through the host's hemocoel and abdomen, absorbing fat cells and other tissues. About 28 hours after infection, most of the *E. muscae* hyphae inside a house fly are clustered around blood cells near the insect's heart. In fruit flies, after 48 hours have elapsed, fungal cells appear around its brain and ventral nerve cord—the insect equivalent of a spinal cord in vertebrates. But while its presence there displaces neuronal cells, the fungus is not killing or consuming host brain tissue (at least, not yet).

Meanwhile, hyphae continue to send questing tendrils into the fly's hemolymph and gobble up its fat reserves. Once those are gone—sometimes within a day or two—the fungus requires one more thing from its host: transportation, to a spot where it can sporulate (spray spores, spectacularly) to entrap the next fly victims. To do that, the fungus begins actively manipulating the fly's behavior. Exactly how this happens is still unknown, though at this point *E. muscae* hyphae have already become established in the fly's nervous system and brain. Time of day appears to be a trigger for the fungus to initiate zombification, ensuring that the fly's death takes place around sunset in preparation for nighttime spore dispersal.

How long a fly lives after *E. muscae* infection sets in can vary; it can be as little as two days or more than ten, depending on the fly species, spore load, and environmental conditions such as temperature and humidity. House flies typically die within five to seven days after being infected, while fruit flies last just four to five days.

A telltale indication that an infected fly's end is near is that its overall activity declines dramatically, signaling that its demise is roughly 36 hours away, Elya and researchers from UC Berkeley reported in 2018 in the journal *eLife* (Elya was at the time conducting her graduate work at Berkeley).[9] In order to observe and record stages of *E. muscae* infection step-by-step, the scientists cultivated a strain of the fungus from an infected wild fruit fly cadaver (*Drosophila hydei*) that they found at a nearby field site. They then used the lab-developed fungus strain to infect *Drosophila melanogaster* fruit flies (which are commonly used in lab research), as well as house flies. Soon after the initial decline in their overall movement, the flies stopped flying, exhibiting "the first portent of imminent death," Elya and her coauthors wrote. These zombie flies could still walk, and they would move if the scientists gently prodded them with the tip of a paintbrush; but flight was off the table.

The zombies' next order of business was to climb—an affliction the scientists referred to as "summit disease"—and the flies climbed for about as long as they were able to walk at all. In the wild, infected flies typically clamber up grasses and other plants or human-made structures above ground level, but the flies in the UC Berkeley experiments climbed wooden dowels provided by the scientists. In most cases, the flies moved normally, though a small percentage grew increasingly unsteady and shaky as they climbed. Nevertheless, even in the most unsteady flies, the climbing persisted.

"We think this is adaptive for the fungus," Elya says. "The fungus positions the fly in this elevated way with this very specific posture, and then grows out through the corpse to spread into the environment. By having the animal stuck up high after it's dead, the fungus is now at this vantage point where it can shoot spores into the environment, and they can drift on wind currents, so they could potentially hit a wider area." The spores are viable for just a few hours, so positioning the fly for optimum spore dispersal is therefore critical for the fungus's reproductive success.

As soon as a fly was done walking, it would extend its proboscis—the elongated mouth-tube that hoovers up food—as drops of a clear liquid oozed from the tip. The liquid (which may be a fungal secretion, though this is yet to be confirmed) is a strong adhesive; when a fly's proboscis touched one of the dowels it was climbing, the insect stuck there so tightly that it couldn't yank itself free (as some of them weakly tried to do). In fact, other researchers had previously noted that this adhesive—whatever it was—was unusually strong, and their attempts to dislodge a dead, stuck house fly frequently resulted in the fly's body breaking off, while the proboscis stayed firmly glued to the climbing surface.

Over the next ten minutes after depositing this sticky kiss of doom, the dying fruit fly would begin raising its wings and spreading them away from its abdomen. This movement didn't happen all at once; rather, the wings would lift briefly, halt, and then lift again in short spurts of motion that was "reminiscent of the inflation of a balloon," until the wings reached their full extension, the UC Berkeley researchers wrote. By then, any body movements, leg twitches, or other feeble signs of life in the fly had sputtered out, and conidiophores began to push through the fly's abdomen. In flies that were examined within two to four hours after they expired, their muscles were still intact, but the fungus had spread throughout their entire bodies and consumed their guts and reproductive organs. What's more, there was "clear evidence" of the fungus in the flies' brains, and the UC Berkeley scientists recorded signs of gene expression (when genetic information is translated into a function) in those fungal cells. Those genes were linked to biological activity in the fungus rather than behavior manipulation of its host. Nonetheless, "their presence in the *D. melanogaster* brain provides various means by which the fungus could communicate with neurons in their own chemical language," the *eLife* study authors reported.

That said, extension of a fly's proboscis and fanning of its wings may be the result of mechanical manipulation rather than chemical puppeteering, Elya and de Fine Licht later wrote in *IMA Fungus*. Physical pressure from the growing mass of hyphae inside a fruit fly's body could force the insect to extend its straw-like snout, just as prior experiments by other researchers showed that a house fly's proboscis pops out when its body is injected with a high volume of fluids. Weighty masses of hyphae could likewise push the fly's wings into a raised position by filling up its abdomen and then pressing down on the wing muscles, causing them to contract.

There's an unexpected twist to a fly's reaching that final form, with wings spread wide over a spore-swollen abdomen: attractiveness to the opposite sex. Perhaps you wouldn't expect a zombie cadaver to broadcast sex appeal. Then again, you're not a male house fly—they seemingly find these dead, bloated, fungus-filled flies irresistible. A male fly that approaches a dead, infected female might pick up adhesive conidia from the halo surrounding his deceased love interest, or he could collect fungal spores during a mating attempt. Even the slightest pressure on an infected cadaver can trigger a spray of conidia from a touch-sensitive, spore-launching cannon.[10]

But male flies seem unaware of or unconcerned with the potential danger of approaching a dead, fungus-filled female, and they even seem to prefer zombie cadavers to living females. In experiments where male flies were presented with a choice of potential mates—a healthy female or an *E. muscae*-infected corpse—they almost always picked the corpse, according to ecologist Anders Pape Møller, a senior research scientist at the Université Pierre et Marie Curie in Paris.

In 1988, Møller decided to investigate whether *E. muscae* manipulated flies through sexual attraction to increase the reach of their spores (at the time, Møller was a researcher in the Department of Zoology at Uppsala University in Sweden). In healthy female flies, a large abdomen signals to males that the female is fertile and carrying lots of eggs. But flies that are artificially inflated with loads of *E. muscae* fungus sport posteriors even portlier than those of the most fertile females—about 30 percent bigger, Møller found. If males are more likely to gravitate toward hefty mates, that innate attraction might make them overlook the fact that a target of their affection was perhaps not a fertile female after all (and was also slightly dead).

To test the attractiveness of a fungus-filled abdomen to male flies, Møller staged experiments that presented males with two options: healthy female or infected corpse. About 90 percent of the males in his trials approached a dead fly first. Møller then went a step further in his tests; scalpel in hand, he frankensteined a group of dead flies by swapping their abdomens, supergluing infected abdomens to the bodies of uninfected flies, and vice versa. Yet again, the flies with bloated fungus-filled abdomens were the healthy males' first picks—regardless of whether or not those abdomens originally belonged to someone else.

Most of Møller's male flies that copulated with the infected corpses became infected too, demonstrating that this sexy strategy helped *E. muscae* spread its spores to new hosts. By altering the appearance of corpses, the fungus manipulated sexual behavior in male flies "to its own benefit," Møller wrote.[11] And even when the abdomens of the dead flies—infected and uninfected—were roughly the same size, flies that were riddled with *E. muscae* were still the first choice for healthy males looking for love.

"This result suggests that features other than the size of the swollen abdomen enhanced the attractivity of infected flies," he reported. Perhaps the fungus wasn't quite finished with its manipulative tricks, even if its original host was already dead.

Møller wasn't the only scientist who suspected that *E. muscae* had a secret chemical arsenal that it was using to attract new victims, ensnaring them with the promise of a risqué rendezvous. In 2022, de Fine Licht and other scientists from Denmark and Sweden found that while the sight of an ample abdomen may contribute to a dead fly's attractiveness, the fungus also uses chemistry to enhance a female corpse's appeal.

When house flies mate, they rely on both chemical and visual cues to locate a partner. The researchers suspected that *E. muscae* was exploiting males' natural reliance on chemical signals by either amplifying the sex pheromones of female flies or by generating an entirely new group of volatiles—compounds that quickly evaporate—that female flies don't normally produce but that still attract males.

It turned out to be a little bit of both.

As in Møller's experiments, male flies observed by de Fine Licht and his colleagues made more attempts to mate with infected female flies than with noninfected females. The males were even more eager to mate if the fungus in female cadavers was in the late stage of spore formation. After these trysts, about 73 percent of the males showed signs of infection.

Was the fungus altering the scent of the corpses to boost their attraction? The scientists tested that by measuring sensory input from the male flies' antennae into their brains using an imaging technique called electroantennography (EAG), finding that the flies responded strongly to puffs of air with *E. muscae* conidia in them. Chemical analysis of the dead flies' cuticles and of volatiles produced by the corpses revealed ingredients for a seductive love potion produced after the flies' deaths. It included compounds made of hydrogen and carbon called alkanes; these are normally produced by female flies and are associated with increased sexual activity in male house flies. Also in the mix were 22 compounds that were typically not found in flies at all. Most abundant among these were sesquiterpenes, another type of hydrocarbon that's common in plants and fungi. And when the scientists examined genetic activity in *E. muscae* colonizing the fly corpses, they found that the fungus was expressing enzymes linked to sesquiterpene production.

One by one, the pieces dropped into place. Male flies were attracted to infected corpses because love was literally in the air, in the form of enticing chemical cues, which the males detected with their antennae. The scientists then linked the production of those male-luring volatiles to *E. muscae* by identifying genetic pathways in the fungus that activated the compounds. *E. muscae* didn't just turn infected hosts into zombie flies—the fungus was also bending the will of uninfected flies.

"The longer a female fly has been dead, the more alluring it becomes to males. This is because the number of fungal spores increases with time, which enhances the seductive fragrances," de Fine Licht said.[12] "Our observations suggest that this is a very deliberate strategy for the fungus," he added. "It is a true master of manipulation—and this is incredibly fascinating." What's even more fascinating is that their investigation provided one of the first descriptions of a mind-manipulating pathogen extending its reach from the corpse of its original host "to also manipulate the behaviour of healthy individuals," the scientists reported in the *ISME Journal*.[13]

*

By attracting healthy males with an airborne love potion produced in the dead body of a former host, *E. muscae* demonstrates an unusual feat for a zombifier: changing insects' behavior prior to infecting them. But it's not the only fungal manipulator that can do this. Another fungus in the Entomophthoraceae family is known for exerting a similar influence over goldenrod soldier beetles (*Chauliognathus pensylvanicus*) and margined soldier beetles (*Chauliognathus marginatus*), inserting itself into the insects' mating dance to increase its chances of spreading infection and zombification.

Meet the beetle-consuming fungus *Eryniopsis lampyridarum*. If you'd been formally introduced to it in 1888, when it was described by American entomologist and mycologist Roland Thaxter, you would have known it as *Empusa lampyridarum*; it was renamed *Entomophthora lampyridarum* in 1891 and then assigned to the genus *Eryniopsis* in 1984. Its soldier beetle targets, also known as leatherwings, are closely related beetle species that are widespread across eastern North America. They are slender, yellowish-orange beetles with elongated abdomens decorated with black markings on their wing cases. Measuring no more than 0.6 inches (1.5 centimeters) long, soldier beetles somewhat resemble fireflies but lack light-producing organs, and they've been around since at least the Cretaceous, based on evidence from fossil specimens preserved in amber.[14]

Thaxter encountered a number of fungi in the Entomophthoraceae family—including *E. lampyridarum*—in insects that he collected in New England and North Carolina in 1886 and 1887. He was hoping to investigate arthropod life cycles but was deeply frustrated when his work was disrupted by pathogenic fungi that selfishly insisted on infecting and killing the insects that he was trying to study.

"I was often greatly annoyed by losing large numbers of larvae and pupae through the agency of fungi," he wrote in 1888 in *Memoirs of the Boston Society of Natural History.*[15]

Thaxter's introduction to *E. lampyridarum* infection in *C. pensylvanicus*, or goldenrod soldier beetles, took place in the fields of Cullowhee, a town in North Carolina. When nearing the end of their lives, infected beetles—like flies zombified by *E. muscae*—displayed signs of summiting disease, climbing upward on flowering plants and then biting down with their mandibles. In the North Carolina fields, Thaxter found numerous dead beetles that the fungus had literally left hanging. The deceased beetles' jaws were clamped so tightly to their perches, Thaxter recalled, that in one field he discovered a number of plants studded with tiny beetle heads "from which the bodies had been broken away." He selected a couple of fungus-invaded corpses and took them back to his lab for further study, but the cadavers were at that point very degraded. His observations focused mostly on the fungus's spores; he collected no data about how the fungus affected living beetles.

After Thaxter published his brief paper on *E. lampyridarum* (along with his unsettling description of gently waving greenery tipped with bodiless zombie beetle heads), the fungus made few appearances in scientific literature over the next century. It resurfaced in 1911, when American entomologists Charles Holcomb Popenoe and Ellison Adger Smyth Jr. described seeing dozens of beetles felled by a fungus, similar to the one described by Thaxter, in Diamond Springs, Virginia, in September of 1909.

"The fungus seemed to attack first the abdomen of the adult, distending it abnormally and producing white, greenish or grayish rings of dense mycelial growth," they wrote. "Often six or more beetles had attached themselves in their death struggle to a single small head of flowers." In death, all of the beetles had assumed "a peculiar posture," with the body positioned upward at a 45-degree angle, the wings held aloft as though the insect were mid-flight, "and the mandibles firmly fastened into the calyx [the protective layer around flower petals] in a last grim death grip." The scientists also remarked that during the previous year in June, Popenoe had spotted thousands of similarly infected beetles covering a blooming dwarf chestnut tree in nearby Norfolk, Virginia. "On closer examination the fact was disclosed that all were either dead or dying from the attacks of a fungus," the authors wrote. They concluded that the fungus in both cases was a near-perfect match to the one that Thaxter had identified in 1888.[16]

Thaxter had also remarked on *E. lampyridarum*'s "very peculiar" secondary conidia—unlike the bell-shaped primary conidia, the secondary spores are watermelon-shaped—and those oddball spores earned a mention in 1973 in the journal *Mycologia*, in a roundup of *Entomophthora* species.[17] But the first allusion to the fungus manipulating beetle behavior didn't appear until 1980, in a paper by Gerald R. Carner, an entomologist at Clemson University in North Carolina (Carner is now professor emeritus).

Carner suggested that the fungus compelled the beetles to grip the plants before they died, an act that provided "a definite advantage to the fungus" because it held the beetles in position while the fungus released its spores, increasing the likelihood of infecting other beetles nearby. The corpses' spread wings were also helpful to the fungus; were the wings not wide open, the dispersal of spores that burst through the beetles' abdomens "would be reduced considerably," Carner wrote.[18]

It was nearly four decades later when entomologists published the first detailed description of the fungal infection in goldenrod soldier beetles based on observations of 446 beetles, dead and alive. Donald Steinkraus, an entomology professor (now retired) from the University of Arkansas, collected the beetles from several locations in Arkansas between the months of September and October in 1996, 2001, 2015, and 2016.

For Steinkraus, zombified soldier beetles were a common sight in Arkansas; he would find them every fall, clinging to plants and covered with spores. He recalls that he started working with the beetles "strictly as a side project" to his university research, which was focused on fungi that sickened crop insects, such as cotton-devouring aphids.

"We developed techniques so that we could predict if epizootics or epidemics in the aphids would take out the aphid populations," he says. But because soldier beetles weren't agricultural pests, there was little interest—and even less funding—for investigating the fungal pathogens infecting them. So Steinkraus, who had long been fascinated by this pathogen-beetle relationship, had to fit that in around his other research. He collaborated in his investigation with Ann Hajek and Jim Liebherr, entomology professors at Cornell University, and they published their results in 2017 in the *Journal of Invertebrate Pathology*.[19]

Of the 446 soldier beetles, about 20 percent were found to be infected with *E. lampyridarum*. Around 19 percent of the males and 21 percent of the females were hosting either conidia or resting spores but never both at the same time. Infected beetles were found on flowers where the insects typically gather in large groups to mate, a phenomenon known as a lek (see figure 5).

"Shortly before death, by unknown mechanisms, dying infected beetles tightly clamped their mandibles into flower heads," Steinkraus, Hajek, and Liebherr reported. This usually took place during the early morning hours, and while healthy beetles typically hold onto plants with their feet, photographs of the zombie beetles showed that their mandibles were the only body part attached to the flowers. About 15 to 22 hours later, between midnight and 7:00 a.m., "the fungus caused dead beetles to raise their elytra [wing cases]" and spread their wings, as the

Figure 5 Goldenrod soldier beetles (*C. pensylvanicus*) infected with *E. lampyridarum* were found clinging to flowers with their mandibles. Their wings were spread wide and bands of fungal spores are visible through their exoskeletons. *Courtesy of D. Steinkraus*

growing fungal masses inside the insects pressed on the wing muscles and erupted through the membranes of the host's abdomen and thorax.

"It'd be as if you brought a body to the morgue, and then in the early hours—like 3:00 a.m.—the mortician walked in and all the cadavers were sitting up," Steinkraus says.

Dead, infected beetles were found clustered together atop flowers representing several local plant species, including the flowering frost aster (*Symphyotrichum pilosum*) and Canada goldenrod (*Solidago canadensis*). Live males that gathered on the flowers to mate eagerly attempted copulation with dead, infected females.

Unlike the fly-destroying *E. muscae* (and most of its cousins in the *Entomophthora* genus), *E. lampyridarum* lacked an explosive finish to its zombifying cycle. None of the infected cadavers were seen forcibly expelling conidia, and the scientists found no traces of ejected spores dusting the wings or bodies of the dead beetles. One possible explanation, they suggested, was that the fungus primarily spread to new victims through direct contact, when healthy insects tried to mate with corpses. To that end, the spread wings of the dead beetles would help to trick prospective mates into approaching and mounting a lifeless insect, by simulating a common mating posture.

This flower-gripping and wing-spreading was only seen in infected beetles that produced the highly transmissible conidia, but the outcome when resting spores were produced was quite different, Steinkraus, Hajek, and Liebherr wrote. "In contrast to the conidial stage of *E. lampyridarum*, infected soldier beetles producing resting spores were not attached to flowers by their mandibles, did not spread their wings, and did not form external conidiophore bands." Rather, those beetle corpses were found lying on the ground under flowering plants. The bodies were brittle and crumbled easily, "and their abdomens were packed with dark brown spherical resting spores." As their bodies fell apart, the spores would drop into the soil, where they would bide their time through the cold winter months, "germinating the next season to infect a new generation of soldier beetles," the researchers wrote.

For beetles infected with *E. lampyridarum* conidia, physical contact is necessary for the spores to infect new hosts. It therefore would benefit the fungus to compel infected beetles to perform actions that mimic mating behavior—climbing flowers, attaching themselves there, and then spreading their wings—to ensure that other beetles will venture close enough for spores to stick to their bodies and infect them, too. However, there is still much to be discovered about how the fungus manipulates its beetle hosts into climbing and latching on to flowers, and whether it releases volatiles (as *E. muscae* does) to lure healthy insects to their eventual doom.

And though Steinkraus is now retired and no longer actively researching this zombie-host relationship, many questions linger for him as well. Some of those involve soldier beetle biology; little is known about the earlier stages of their life cycle and their vulnerability to the fungus as eggs and larvae. But most of his questions are about the fungus.

"I would like to know what the host range is," Steinkraus says. "We don't really know anything about the resting spores. We don't know anything about how the spores are entering the host. There's tons that's not known about this pathogen. Tons."

*

One thing that scientists *do* know is that, fortunately for humans, most fungi in the Entomophthorales order don't extend their mind-controlling mycelium tendrils into any offshoots of the mammal family tree. (There are two Entomophthorales genera—*Conidiobolus* and *Basidiobolus*—that are known to be pathogenic in humans, but these fungal infections are extremely rare. They typically cause swelling in the skin and subcutaneous tissues, and the fungi don't alter host behavior.)[20]

But other types of noninsect hosts aren't so lucky.

Even as scientists unlock the secrets of *Entomophthora* insect-destroyers, the well-known zombifiers likely have many more undescribed cousins; to date, researchers have explored only a minuscule percentage of habitats where *Entomophthora* may hold sway. As Elya and de Fine Licht wrote about the genus: "As more scientists become aware of these fungi and more of *Entomophthora*'s potential range is probed, we expect to find additional species."

In fact, a previously unknown genus and species of insect-destroying fungus was recently found to have taken an evolutionary side step to infect and manipulate a noninsect type of arthropod: millipedes.

Sometimes called thousand-legged worms, millipedes have long, cylindrical bodies made up of numerous rounded segments, each with two pairs of legs. There are about 12,000 known species worldwide (and about 80,000 estimated species), and they dwell in soil in moist terrestrial habitats where many insects live. Yet millipedes are not insects. Nor are they worms, for that matter; they're more closely related to lobsters, crayfish, and shrimp. Most of the time, millipedes hunker down in damp earth and burrow under leaf litter, rocks, and fallen trees. But since the nineteenth century, scientists have reported finding millipede corpses out in the open and perched in high places where the leggy arthropods typically never go. And recent analysis suggests that a zombifying Entomophthorales fungus is to blame.

Like the fly-destroyer *E. muscae* and the beetle-busting *E. lampyridarum*, the arthropod-eating fungus *Arthrophaga myriapodina* also compels pre-death climbing in its zombie millipede hosts, presumably to aid in spore dispersal. It's the first Entomophthorales fungus known to infect millipedes, researchers reported in 2017 in the *Journal of Invertebrate Pathology*.[21] And while it didn't receive a scientific name until the twenty-first century, its stranded, mind-manipulated millipede victims have been intriguing biologists for more than a hundred years.

For the 2017 study, Ann Hajek and mycologists Kathie Hodge at Cornell University and Andrii Gryganskyi at Duke University, conducted molecular analysis of DNA extracted from fungus tucked inside infected millipedes. Observations and measurements of conidia shapes and other fungal structures also helped them identify the fungus as a new genus and species that was closely related to fungus in the genera *Eryniopsis*, *Entomophaga* and *Entomophthora*. The newly described species's name, *A. myriapodina*, refers to the subphylum Myriapoda, which includes millipedes and centipedes. The researchers identified the fungus from sampling three types of brightly colored (and toxic) infected millipedes: undescribed species in the *Nannaria* genus, the millipede species *Boraria infesta*, and the cherry-almond-scented millipede *Apheloria virginiensis corrugata* (its delicious aroma comes from defensive cyanide compounds, which it can produce in amounts potent enough to kill six mice).[22]

Infected millipede corpses in the wild "were typically found in elevated locations, such as the highest point on a decaying tree lying horizontally on the ground," the scientists wrote. In fact, dead *A. virginiensis corrugata* zombies were a common sight in forests around Ithaca, New York, and were often found in locations that were peculiar for burrowing millipedes: "on cement bridge abutments, on top of fence posts as high as shoulder level, and on the elevated ends of fallen branches and logs." Once a host was dead, branching fungal filaments and conidiophores pushed through body segment seams from inside, rupturing the cuticle. The millipedes' exposed positions in high places were probably the result of fungal manipulation, "as the elevated location likely enhances dispersal of conidia ejected from the cadaver," the researchers reported.

Millipede specimens were all collected from sites in northeastern North America, and the study authors dug deep into two dozen entomology collections that spanned 130 years, to find the earliest observations of the relationship between the fungus and its millipede hosts. They then compared samples of the fungus from freshly collected millipedes and from collection specimens that were dried or preserved in alcohol.

The oldest example, acquired in 1886 by Roland Thaxter, was a *B. infesta* specimen that was bursting at the body seams with fungus. Though Thaxter had not described the millipede's killer, when Hodge, Hajek, and Gryganskyi examined Thaxter's millipede, they confirmed that the fungus matched their modern samples of the newly described *A. myriapodina*. Thaxter also received samples of fungus-infected millipedes in 1916 from A. T. Speare, a fellow biologist with the US Department of Agriculture; those specimens are now part of the Farlow Herbarium collection at Harvard University. Thaxter labeled and stored them as millipedes infected with "*Entomophthora myriapodina*," but the fungus was never formally described under that name. In the new study, these specimens were also identified as examples of *A. myriapodina.*

Hodge, Hajek, and Gryganskyi then consulted a modern resource that was unavailable to Speare and Thaxter: the internet. They examined photos posted online by multiple sources that showed dead, stranded millipedes seemingly infected with a fungus that resembled *A. myriapodina*. By geolocating the photos, the researchers were able to roughly map an even broader potential range for the newly described zombifying fungus. It extended over New York, Maryland, Virginia, and Washington, DC in the United States, and across the southern part of Ontario in Canada. Other online images suggested that the fungus might also be thriving as far south and west as Texas and California.

But the researchers were unable to replicate conditions in the lab where they could observe *A. myriapodina* fungal infection in a millipede host from start to finish. And though the height-seeking, pre-death climbing of the millipedes corresponds with similar behavior in other zombified arthropods, the mechanisms of the fungus's mind manipulation remain hidden. While some aspects of this fungus are now more clearly defined, "further research is needed to understand the disease cycle of *A. myriapodina*," the scientists wrote.

However, there may be a ray of light at the end of the tunnel for understanding what drives climbing in fungus-manipulated zombies, more than 150 years after zombie flies first captured scientists' attention.

In 2023, Elya and a team of researchers at Harvard University unlocked some of those summiting mechanisms in fruit flies (*Drosophila melanogaster*) infected with *E. muscae*, using the laboratory model that Elya had developed while at UC Berkeley (the scientists dubbed it "the zombie fruit fly system").[23] This system, in which a lab-cultivated strain of *E. muscae* was introduced into fruit flies, enabled the researchers to not only zombify fruit flies at will but also to cultivate the fungus outside of a fly host so they could more closely observe how *E. muscae* grew and behaved, using artificial environments that mimicked the nurturing innards of a fruit fly.

And because fruit flies are a model organism—a laboratory staple used in many types of research—Elya and her coauthors already had ample data at their fingertips about how fruit fly brains were wired and how they worked. This would be essential for identifying the unique fingerprints of *E. muscae* infection in the brains of their zombies.

"Scientists have studied the fruit fly for such a long time and developed so many tools that the number of things we can do in flies is just astronomical," Elya told the *Harvard Gazette*. "It was the ideal system to understand the mechanistic basis of how the fungus manipulates behavior."[24]

Their first steps were training a computer with behavioral datasets and then using machine learning to automate the tracking of hundreds of infected flies in the lab, to identify summiting as it happened in real time. During their observations of summiting behavior, the researchers realized that flies demonstrated a burst of speedy locomotion about two and a half hours before they died. And if there was a surface for them to climb, that energy carried them upward.

The next goal was finding which parts of the flies' brains were activated during summiting, and which neurons were involved. When the study authors peered at images of the fly brains, they found that the fungus displaced brain tissue, creating pockets where it could settle and grow. But it didn't consume those tissues or degrade neurons or brain structures until the fly had climbed as high as it possibly could and then died. By this time, other organs inside the fly were "essentially obliterated," the scientists wrote. The brain may have been intact, but the fungus was firmly in control. Brain tissue analysis of summiting flies showed that the fungus was targeting a brain region known as the pars intercerebralis (the fruit fly equivalent of the hypothalamus), involving neurons that regulated circadian rhythms and secretions of juvenile hormones related to metamorphosis. Activating these neurons in uninfected flies turned them into climbers, and disabling or mutating the neurons reduced summiting—and so did blocking the release of juvenile hormones, the scientists discovered.

Pathogens in the blood are typically prevented from reaching the brain by a layer of cells known as the blood-brain barrier, which isolates and protects the brain. But in the case of the infected flies, *E. muscae* weakened that barrier. A fly's hemolymph was then able to bathe its brain with fungal payloads, activating neuron activity to trigger climbing. The researchers also found that when they drew hemolymph from infected, summiting flies and injected it into healthy flies, the transfusion would cause accelerated locomotion in the flies they injected. Weakening the blood-brain barrier could be a key part of the fungus's strategy for escalating its internal takeover of the fly and progressing from feeding on its body tissues to manipulating its neural activity.

"These parasites, they're effectively natural neuroscientists," Elya says. "Even the best neuroscientists among us would be hard-pressed to be able to manipulate the behavior of their organism of choice in such dramatic ways."

Of course, zombifiers like *E. muscae* have had millions of years of evolution to crack the code of insects' central nervous systems and compel unwilling hosts to help them reproduce. If scientists can figure out exactly how zombifiers accomplish that manipulation, Elya says, "we'll have much more insight into how nervous systems generate behavior, and how they can be perturbed in order to change behavioral outcomes."

4

THE SLEEPER AWAKENS

Zombie Cicadas

"Is he alive or dead? Well, that's the question nowadays isn't it?"

—DR. LOGAN (RICHARD LIBERTY), *Day of the Dead*

I met my first zombie on May 23, 2021. I was at New Jersey's Princeton Battlefield State Park, against a backdrop of unspeakable carnage, in a field littered with tens of thousands of corpses. It was nearly impossible to avoid stepping on them, and in the stillness of the sunny spring afternoon under blue skies, every step was accentuated by the crunching sound of bodies splintering underfoot. Piles of scattered remains littered the grass and were densely clustered under trees, and empty husks clung to branches and dangled from leaves that they clutched tightly in a lifeless grasp.

The zombie blundered into me quite suddenly and unexpectedly and started exploring my leg. It was somewhat less terrifying than the typical ravening movie ghoul because it measured only one inch long—smaller than my thumb. While I recognized it as a zombie, I knew I was safe from being bitten because its tiny mouth couldn't pierce human skin. And though we were surrounded by corpses, the bodies littering the grass weren't the aftermath of a zombie plague. Though there were a few zombies among them, most of the dead represented a natural stage in the life cycle of periodical cicadas—flying insects in the *Magicicada* genus that burst from the ground during springtime, after living beneath the soil as immature nymphs for more than a decade. They emerge in vast armies numbering in the billions, and in some spots may appear in concentrations as dense as 1.5 million insects per acre.

Once aboveground, the nymphs climb the nearest tree (or a tall blade of grass or a wall if no trees are nearby), wriggle out of their exoskeletons and transform into black-bodied, red-eyed adults. They live for around four to six weeks, devoting their precious time to bouts of raucous singing and mating. Eventually the females lay 400 to 500 eggs—usually not too far from the spot where they themselves emerged—inserting them into tree branches in slits that they punch with a sharp ovipositor. The adults die, and the eggs hatch about six to ten weeks later. Newly hatched nymphs fall to the ground and burrow into the soil, and then the cycle begins again as they settle in for a leisurely subterranean childhood, which can last for up to 17 years, depending on the cicada species.

But the cicadas aren't alone during this prolonged nymph stage. Languishing alongside the buried juveniles is a patient and deadly enemy: a pathogenic fungus named *Massospora cicadina*. The choreography of *M. cicadina*'s life cycle matches its steps to the cicadas' unusual biological clock in a deadly pas de deux; as the years tick by, it lies in wait, only to start its zombifying work when its cicada hosts emerge from deep underground and begin their metamorphosis into adults. Other *Massospora* fungi target annual cicadas and have spores that rest for two to five years until their potential hosts crawl from the ground, but *M. cicadina* only affects periodical cicadas, which emerge on a 13-year or 17-year cycle. It's the only known pathogen—or predator of any kind, for that matter—that targets periodical cicadas by synchronizing its own life cycle to that of its unusually long-lived insect prey.

The zombie cicada that found me in Princeton was an adult in a highly visible stage of a deadly fungal infection; while the upper third of its body was a deep midnight black, the lower third was a vivid yellow. Its exoskeleton around the abdomen had burst open, and in its place was a plug of tightly packed fungal spores that were primed for dispersing and infecting other cicadas. Yet even though this partially consumed cicada was missing a substantial chunk of its body, it was still very much alive—though literally a fraction of its former self.

Our paths crossed in that open field as the zombified insect was on the hunt—not for brains, as a fictional human zombie might be, but for mates. The cicada's fungal hitchhiker was compelling its host to reproduce, even though the insect's reproductive organs had already dissolved and the rest of its abdomen was falling apart. Periodical cicadas spend most of their adult life stage focused on reproduction, but in this case the insect's normal mating urge had been kicked into overdrive by the parasitic fungus. An array of chemicals, some of which included psychoactive compounds, were coursing through what remained of the insect's body, driving the deteriorating cicada to complete one vitally important task: mate with any cicadas it could find. Male cicadas typically sing to attract females, vibrating special corrugated exoskeleton membranes on their torsos, called tymbals, to produce their distinctive calls, to which females respond with a staccato clicking sound that they make with their wings. But infected males under the fungus's control don't just seek females. They also respond to other males' calls by clicking as females do, enticing the eagerly calling males to come and mate with them. In doing so, zombified males help *M. cicadina* accomplish something the fungus couldn't do alone: spread its spores among as many cicadas as possible—male and female alike.

The relationship between *M. cicadina* and its periodical cicada victims is a unique tale about the evolution of zombifying superpowers and how a zombifier's own life cycle can evolve over time in tandem with one species of victim; in this case, the zombifying *M. cicadina* is perfectly synchronized to match the cicadas' unusually lengthy underground life stage. And unlike other types of zombifying fungus that modify host behavior but hold off on jettisoning their spores until their insect hosts are corpses, *M. cicadina* takes a different approach. Instead of waiting until an infected cicada is dead, the fungus compels its hosts to start actively distributing spores while the maimed and fungus-stuffed insects are still alive, a strategy known as active host transmission. Zombified cicadas may infect others by seeding the air and ground with showers of spores that they shed from their bottoms, a behavior that in 2013 led Cornell University student Angie Macias, now a researcher in plant and soil sciences at West Virginia University, to call the insects "flying salt shakers of death" in a post that she wrote for the Cornell Mushroom Blog.[1] Cicada zombies also pass the infection along by mating with uninfected cicadas and transferring spores directly onto their partners' genitalia, in a ghastly act of sexually transmitted zombification.

There are more than 3,000 cicada species, but only seven of these are periodical cicadas. With midnight-black bodies, orange wing veins, and vivid carmine eyes, periodical cicadas look strikingly different from annual cicadas, which are typically greenish and patterned with brown or dark gray, and have green or black eyes. Annual cicadas are common around the world, emerging to sing during the summer months (typically from July through early September in the Northern Hemisphere). But periodical cicadas are found only in the eastern United States and emerge in springtime, usually from May through June. They've been spotted as far west as Oklahoma, as far north as Michigan and New York State, and as far south as Louisiana, and they follow a life cycle that keeps them underground for 13 or 17 years, with four species in southern states following a 13-year cycle, and three species in more northern locations appearing every 17 years. Multiple periodical cicada populations across more than a dozen US states follow these two cycles, and groups of cicadas that emerge during the same year—regardless of where they pop out of the ground or which species they belong to—are categorized as part of the same "brood."

Periodical cicadas are sometimes mistakenly referred to as "locusts," a swarming, migratory phase in certain types of desert grasshoppers in the taxonomic order Orthoptera, which also includes crickets and katydids. Locusts are typically solitary, but certain environmental factors, particularly those that bring favorable breeding conditions and cause lots of locusts to gather together, lead to physical changes in the insects, causing them to shift into what is known as a gregarious phase. And gregarious locusts are swarming locusts. These swarms, which can travel together as far as 100 miles (161 kilometers) per day, may contain trillions of individuals, as did the so-called plague of locusts that emerged in Eastern Africa in late 2019 and persisted through 2020, producing the worst swarms seen in the region in about 70 years. Locusts can be highly destructive to agriculture, blanketing regions with ravenous insects that devour vegetation and leave acres of gnawed wasteland in their wake. But while the sheer numbers of periodical cicada broods can sometimes appear to rival those of locust swarms, cicadas are part of a different insect order—Hemiptera, or true bugs, which have sucking mouthparts. Cicada brood emergences can number in the billions; but while they may seem to swarm like locusts, they don't migrate as a flying group over great distances. Nor do they devour crops as locusts do; cicada mouthparts are suited for piercing roots and stems and sucking liquids, not chewing up plants.

There are currently 15 active broods of periodical cicadas, each identified by a Roman numeral (broods are numbered from one to 30, but some of those populations are extinct and other formerly designated "broods" are now thought to be straggler cicadas from other brood years). In 1893, Brood I became the first officially named brood of 17-year cicada populations, and scientists identified a new brood each subsequent year up to Brood XVII, and Brood XVIII to Brood XXX refer to 13-year cicadas in southern states. The US Department of Agriculture reports 17-year cicadas in multiple eastern locations dating to the mid-nineteenth century, with sightings in two New Jersey cities—Princeton and Lambertville—dating back to 1800.[2]

The Princeton zombie that I encountered in 2021 was part of Brood X, also known as the Great Eastern Brood, which includes all three 17-year cicada species: *Magicicada septendecim*, *Magicicada cassini*, and *Magicicada septendecula*. Brood X is one of the biggest broods, if you're judging brood size by how widespread its cicada population is, and in 2021 this brood emerged across 15 Mid-Atlantic, Midwestern, and Southern states—Delaware, Georgia, Illinois, Indiana, Kentucky, Maryland, Michigan, New Jersey, New York, North Carolina, Ohio, Pennsylvania, Tennessee, Virginia, and West Virginia—as well as in Washington, DC. Between late April and June that year, billions of cicadas—zombies and nonzombies alike—blanketed trees and covered buildings and sidewalks. The densest concentrations were in Indiana and Maryland, where millions of cicadas popped up statewide.

This particular generation of Brood X hatched in 2004, when Facebook had just launched but was only available to Harvard students, and when George Bush was still the US president. As hatchlings, cicada larvae have white bodies and are about the size of a rice grain, measuring just one-sixteenth of an inch. The freshly hatched nymphs pushed from their eggs, which had been carefully stuffed into slits in tree branches by their mothers, dropped to the ground and immediately burrowed into the soil where they stayed for the next 17 years, sucking a nutritious liquid called xylem from grass and tree roots and molting through five growth stages, or instars. Waiting right alongside them in the dirt were *M. cicadina* spores, patiently biding their time as the years ticked by. Some nymphs would pick up those spores when they were newly hatched, as they burrowed into the ground. Others would acquire spores on their journey to the surface, 17 years later. But only when the nymphs were ready to shed their exoskeletons and complete their metamorphosis into adults could zombification truly begin.

The earliest mention of zombified cicadas—even if the writer didn't call it zombification at the time—is a journal entry written in 1800 by Benjamin Banneker (1731–1806), a free Black man and farmer in Maryland, and a self-taught mathematician and astronomer. From 1791 to 1802, Banneker authored and published the annual journal *Pennsylvania, Delaware, Maryland, and Virginia Almanac and Ephemeris*, documenting his calculations for tidal activity, eclipses, and other astronomical phenomena. He also conducted other observations of the natural world, an accomplishment that was highlighted by John H. B. Latrobe, a lawyer and inventor, when Latrobe read from Banneker's memoir at a monthly meeting of the Maryland Historical Society on May 1, 1845. Though Banneker had no formal training as a naturalist, he proposed in his journal writings that periodical cicadas appeared in a 17-year cycle, based upon his own local sightings of the massed insects in 1749, 1766, and 1783. According to a written record of Latrobe's 1845 reading, Banneker wrote in his memoir that even though cicadas' lives were short, their song-filled days were "merry." Banneker didn't know it at the time, but he also provided the first known report of a zombified cicada, noting in his memoir that in some cases, a cicada's so-called merriment continued uninterrupted even after the lower part of the insect's body had "rotted" completely away. This gruesome development "does not appear to be any pain to them, for they still continue on singing till they die," he wrote.[3]

Banneker may not have identified this peculiar condition as a disease or fungal infection, but other researchers who gravitated to these insects questioned if there was an outside factor that was causing the cicadas' abdominal decay and began investigating it more closely. Scientists got a detailed glimpse of the cicada-dissolving pathogen in 1851, when Joseph Leidy, a zoologist, parasitologist, and paleontologist, illustrated "a fungous [*sic*] disease" affecting cicadas, publishing drawings and a description in *Proceedings of the Academy of Natural Sciences of Philadelphia.*[4] Leidy based his work on examinations of more than a dozen infected cicada specimens that he had collected. In several specimens, Leidy wrote, the cicadas' abdomens were "filled with a mass of oval spore-like bodies."

The fungus finally got the name *Massospora cicadina* in 1878, courtesy of mycologist Charles Horton Peck, who formally described it in an annual publication for the New York State Museum of Natural History.[5] As to how infected cicadas spread the fungus to their cicada friends and neighbors, an 1898 report released by the US Department of Agriculture's Division of Entomology offered a vivid description: "Upon breaking off the hind part of the abdomen, the dust-like spores would fly as from a small puffball," researchers wrote in the report. They added that mating attempts by infected cicadas with partial abdomens often ended with the maimed insect's damaged body part separating during sex and left dangling from the abdomen of the healthy partner.

But before *M. cicadina* fungal spores can fly, float, or cling to their victims and begin consuming them from within and triggering zombification, they must wait.

And wait.

And wait.

Months and then years pass as developing periodical cicada nymphs sip from plant roots underground and enjoy the relative peace and quiet of subterranean life. Meanwhile, *M. cicadina* slumbers in the form of resting spores—reproductive structures that are thick-walled and more durable than conidia (the spores that are produced asexually and are the fungus's primary means of dispersal and reproduction). *M. cicadina*'s earliest chance to infect cicada nymphs is when the insects are hatchlings, as they burrow into the ground. But resting spores can also latch onto emerging fifth-instar youngsters when they begin building the vertical burrows through which they will creep to the surface. After the tunnels are constructed, nymphs will crawl through them into the open air a few weeks later, once soil temperatures reach 64 degrees Fahrenheit (18 degrees Celsius) at a depth of about 8 inches (20 centimeters). Should those nymphs encounter resting fungal spores along the way, the fungus collects and settles in the nymph's abdomen. But those spores have to remain patient for a bit longer; a nymph needs to transform into an adult cicada before its fungal hitchhiker can begin its zombifying work.

By the time a cicada nymph is ready to creep from the ground and enter adulthood, it measures about 1 inch (2.5 centimeters) long and is encased in a light brown exoskeleton. However, it won't look like that for long. When nymphs emerge from their buried tunnels—an event that takes place at night, often with hundreds or even many thousands of nymphs appearing at once—they seek a vertical surface to climb; this can be a tree trunk, a shrub, a slender grass stem, or even the wall of a building. As a nymph clings to its perch, its exoskeleton splits down the back. Over the next few hours, a red-eyed adult with a pale, soft body and tightly furled wings will wriggle out of the husk. Its body darkens from vanilla to licorice as the cicada's new exoskeleton hardens, and its orange-veined wings unfurl and dry. After a short rest, the newly minted adult is now ready to fly away to seek a mate—and if the cicada is unfortunate enough to be carrying spores in its abdomen, it's about to undergo a much more unpleasant transformation.

Immediately after metamorphosis there is usually no outward sign that an adult cicada is infected and on a path toward zombification. But inside the insect's body, changes are already underway. In 1976, researchers reported the discovery of an early sign of fungal infection in Brood V cicadas, which had emerged in Ohio in 1965. Though the infected cicadas looked normal, the liquid in insects that is akin to blood—hemolymph—in their abdomens, was milky rather than clear, and it had a ropy texture. Its milky color came from the cicada's abdominal organs, which had started to dissolve, and floating in the fluid were fungal filaments that measured between 15 and 40 microns long. From that milky stage, it took about a week for fungal production to kick into high gear, filling the abdomen with a whitish mass of spores and completely consuming the insect's soft tissues. As the mass grew, it pressed on the abdominal walls, weakening connections between segments until the walls ruptured and the abdomen dropped away piece by piece, leaving the still-living cicada with no reproductive organs and an exposed yellow-white spore plug where its rump used to be.[6]

An adult periodical cicada that was infected with *M. cicadina* resting spores as a nymph develops what is known as a Stage I infection. As the fungus germinates, it produces conidia spores that collect together in a tightly packed clump, expanding and then eventually rupturing the infected insect's abdomen. When these conidia transfer to other adult cicadas, usually through direct contact during mating attempts, they create a Stage II infection in which the fungus reproduces inside the new host as resting spores. As resting spores multiply and accumulate, they also eventually crack and then shatter the host cicada's abdominal wall. But resting spores are more loosely packed than conidia, and they fill the abdomen with a mass of loose, dry spores. It isn't hard to imagine such spore-filled cicadas as Macias's "flying salt shakers of death," with the insects flitting from tree to tree carrying their deadly abdominal payloads and showering the ground—and other cicadas—with zombification and death from above. As these broken, infected Stage II cicadas continue to crawl and fly, resting spores scatter from their bodies and settle on soil where they may be picked up by still-emerging nymphs or by new hatchlings on their way underground. They may also languish in the soil for 13 to 17 years, then attach themselves to emerging fifth-instar nymphs and begin their zombifying work when the nymphs are ready to transform.

In 1983, scientists compared the behaviors of cicadas with Stage I infection to cicadas with Stage II infection; they examined Brood VIII cicadas in Ohio, which emerged in 1968; and Brood V cicadas in Pennsylvania, which appeared in 1982. The researchers didn't just observe cicada behavior in and around the forests where the insects emerged; they also collected Brood V cicadas to record data about flight length in both infected and uninfected individuals, conducting flight tests with free-flying cicadas outdoors as well as tethered flights with insects in the lab. Cicadas with Stage I infections spent more time walking, leaving trails of spores behind them, the scientists reported in *Oecologia*.[7] By comparison, cicadas with Stage II infections spent more time in the air, many of them with a visible cascade of loosely packed spores trickling from their bottoms as they flew. Cicadas in both infection stages tended to perform shorter flights than uninfected cicadas and would drag their abdomens on the ground as they walked, leaving spore trails in their wake. While this behavior may have been intended to rid themselves of the fungus, it served a purpose for the fungus, too, dispersing even more spores and spreading them over a wider area.

The distinctive yellowish plugs of Stage I infections are usually more visible in males, says mycologist Matt Kasson, an associate professor of plant pathology and mycology at West Virginia University who studies the cicada-zombifying fungus. In males, "it's really easy for the abdominal sclerites—these rings that make up the exoskeleton—to kind of slough off, one ring at a time. The top ring comes off, and the ring after that." With females, by comparison, the abdominal rings usually don't snap off by themselves because they're supported by muscle tissue connected to the ovipositor. But with enough pressure, her backside is still brittle enough to break.

"Oftentimes, the way you know a female's infected is if she interacts with a healthy male and they try to mate, he'll break off her genitalia and abdomen when he flies away," Kasson says. "It's pretty grotesque."

More than a hundred years ago, Banneker wrote that a maimed cicada scarcely seemed to notice the loss of its abdomen. Within just a few decades of Banneker's observation, the authors of the *Oecologia* study reported that infected cicadas would fly around "in an unsteady manner" or were "sluggish and averse to flight," making them easier to capture. As researchers gradually pieced together the details of how *M. cicadina* infected cicadas and what those infections did to their bodies, more questions arose about the cicadas' behavioral changes. Were they acting differently than they would if they were uninfected as a purely physical response to the severe damage *M. cicadina* caused in their bodies? Or had the fungus also wrested control of their brains as well as their abdomens, and was it actively manipulating its cicada hosts' behavior for its own benefit?

Differences in how infected cicadas walked or flew weren't the only behavioral changes that teams of scientists observed over the years; in 2018, researchers reported that cicadas with visible *M. cicadina* infections displayed unusual sexual behavior, too. Imagine that you're an infected cicada, and a fungal infection has just burst your abdomen and replaced your sexual organs with a spore plug (see figure 6). Given the circumstances, you might expect that reproduction would be the last thing on your mind. However, these plug-sporting cicadas are still quite keen to reproduce and in fact perform very enthusiastically, researchers noted.[8] Infected cicadas' sexual partners also don't seem to care that their partners are missing some pretty important parts.

Figure 6 A "zombie" Brood X periodical cicada with the telltale yellow plug of spores where its butt used to be, found in Princeton, New Jersey, in 2021. *Courtesy of Mindy Weisberger*

"It is relatively common," the scientists wrote in *Scientific Reports*, "to find a healthy cicada with its genitalia plunged into the abdominal spore mass of an infected partner, or to see healthy cicadas attached to fragments of abdomen or terminalia that have torn free from infected partners during attempted copulation."

Infected cicadas also behave in ways that they normally wouldn't in order to attract potential mates, the study authors discovered. Uninfected cicada males typically signal to females by vibrating the tymbal membranes on the sides of their bodies, producing a song that rises to a buzzing crescendo and is unique to that particular species. Receptive females then show their interest with a clicking sound that they make by flicking their wings. People can easily imitate this click by snapping their fingers, as nature documentary host Sir David Attenborough skillfully demonstrated in the 2005 BBC program *Life in the Undergrowth*. He used repeated finger snaps to lure an eager male cicada along a branch, leading it first in one direction and then another. Attenborough's imitation of a female cicada's mating call was so effective that eventually the male flew right onto the documentarian's finger-snapping hand.

For the *Scientific Reports* study, scientists at Wesleyan University's College of Integrative Sciences and at the University of Connecticut's Department of Ecology and Evolutionary Biology used recordings of male cicadas' songs to see how infected cicadas responded. They found that male cicadas with Stage I infections not only vibrated their tymbals to attract females; they also responded to the sounds of nearby calling males by flicking their wings, producing a clicking noise with the same species-specific timing and pitch that a female's wing flicks produced. By broadcasting this signal to interested males and thereby suggesting that a female was close by, diseased insects enticed fungus-free males into mating attempts that exposed them to infection.

"Thus, *Massospora* functions at least partly as a sexually transmitted disease," the study authors reported. They then added that the female-like behaviors of infected males were, in fact, "complex manipulations instigated by the fungus for its own benefit." In other words, when males flicked their wings, the fungus—somehow—was making them do it.

It was only the males with Stage I infections—their abdomens burst and plugged with masses of conidia spores—that displayed this female-imitating behavior. This hinted that however *M. cicadina* was compelling cicadas' behavioral changes, its strategy followed paths specific to both cicada biology and the fungus's own reproductive cycle. The behavioral changes were sex-specific, as the fungus only affected signaling behavior in male cicadas. But these changes also only took hold in males that hosted *M. cica-*

dina's conidia spores. Stage I–infected males flicked their wings as females did, while males with Stage II infections didn't. Whatever *M. cicadina* was doing to alter the males' mating behavior, that mechanism was linked to a specific stage of the fungus's reproduction.

Whenever mycologists want to understand how a fungus accomplishes something bizarre and complex, such as behavior manipulation, one of the first places they look for clues is in secondary metabolites—"small molecules that aren't essential for primary function but are really spectacular when it comes to defense and interactions," says Kasson. But studying those in *M. cicadina* isn't easy because the fungus can't be cultured. Instead, it must be sampled from field-collected infected cicadas or from preserved specimens in museums.

It took Kasson and a team of more than two dozen scientists from institutions in the United States and Denmark more than three years to identify key chemical components that offered a possible explanation for hypersexual behavior in cicadas hosting *Massospora* fungus. Through genetic and chemical analysis of dozens of spore plugs, the researchers found that infected periodical cicadas are likely experiencing an infusion of a potent mind-bending compound: a plant-associated amphetamine, called cathinone. They also learned that other *Massospora* species that target annual cicadas may take a hallucinogenic route to alter cicada behavior, releasing a compound associated with so-called magic mushrooms: psilocybin.

Between 2016 and 2019, the scientists reviewed hundreds of cicada specimens—infected and healthy, freshly gathered and in museum collections—to find clues about how *Massospora* fungus could affect cicadas' sexual behavior. The researchers sampled Brood VII periodical cicadas in New York during the 2018 emergence, and annual wing-banger cicadas (*Platypedia putnami*) from New Mexico and California, in 2017 and 2018, respectively. They also examined infected Brood V cicadas that emerged in West Virginia and Ohio in 2016, Brood VI cicadas from North Carolina in 2017, and another annual species called Say's cicadas (*Okanagana rimosa*) that emerged in Michigan and had been previously collected in the summer of 1998. For additional and older specimens, dating from 1978 to 2002, the scientists turned to the University of Connecticut's collection; and a collection at Mount St. Joseph University in Cincinnati provided fungal plugs from periodical cicadas that had been collected between 1998 and 2008.

From those specimens, the scientists analyzed 51 *Massospora*-infected cicadas. These included 26 periodical cicadas with *M. cicadina* infections, 15 wing-banger cicadas that were infected with *M. platypediae*, and 9 Say's cicadas infected with *M. levispora*. Most of the infections were Stage I, with the insects sporting visible plugs of conidial spores. But 13 of the cicadas showed Stage II infections, with resting spores.

Their findings, published in 2019 in *Fungal Ecology*, identified cathinone in the periodical cicada samples.[9] Cathinone is an alkaloid that previously was known only from the khat plant (*Catha edulis*), a flowering, evergreen shrub that grows in East and Southern Africa and on the Arabian Peninsula. For thousands of years, people in those regions have chewed khat's young shoots and leaves as a stimulant; psychoactive molecules in cathinone are similar to those in amphetamines, accelerating message transmission in the brain. In humans, the plant is consumed "to alleviate fatigue, enhance work capacity, stay alert, reduce hunger, and induce euphoria and self-esteem," a different team of researchers reported in 2021, in a review of scientific literature describing how the drug affected human behavior.[10]

In the *Fungal Ecology* study, the scientists detected cathinone in periodical cicadas with Stage I and Stage II *Massospora* infections, and the levels of cathinone were four to five times higher in cicadas with Stage I conidial spore plugs. By comparison, annual cicadas with *Massospora* infections were cathinone-free. However, they contained a different chemical: the psychoactive compound psilocybin, which is also found in more than 100 species of mushrooms in the *Psilocybe* genus. *Psilocybe* fungi—so-called magic mushrooms—are consumed by people seeking the mushrooms' hallucinogenic effects, a practice that for millennia has been part of religious rituals in Indigenous cultures in Central America and Mexico. When humans ingest psilocybin, the body converts it into psilocin, which can cause a sense of euphoria and heightened emotions; it can also affect the perception of time, and produce visual and auditory hallucinations.

Prior research had already documented changes in infected cicadas' behavior. Now, the *Fungal Ecology* study's findings provided the first clues about the chemistry that might be driving *Massospora*'s mind-control and manipulation of their cicada zombies. In periodical cicadas, amphetamines could help "improve endurance" in badly damaged hosts, the study authors wrote, enabling cicadas to continue seeking partners and mating despite the loss of their sexual organs and abdomens. Other teams of researchers in the 1970s and 1980s had explored amphetamine's impact on other insects, such as ants and blowflies.[11] They discovered that doses of amphetamines at levels comparable to those in infected periodical cicadas increased aggression and reduced interest in feeding.[12] These earlier findings hinted that the cathinone produced by *Massospora* infections in cicadas might cause similar reactions, which could be driving the cicadas' hypersexual behavior.

It's still not clear if the psychoactive compounds in zombie cicadas are produced directly by *Massospora* fungus, or if a *Massospora* infection triggers some type of chemical response in the cicadas, so that the insects themselves produce the mind-altering chemical. A closer look at the DNA of the fungus and the cicadas would help to resolve that question, Kasson and his *Fungal Ecology* coauthors wrote, suggesting that "higher quality genome assemblies of fungi and hosts" would be required.

However, studying periodical cicadas—and their fungus zombifier—presents unique challenges for scientists, as the cicadas' unusually lengthy life cycle makes it impractical to raise and observe them under controlled laboratory conditions. Scientists therefore rely on preserved specimens in research collections, or else they need to time their investigations to synchronize with the emergence of cicada broods every 13 or 17 years, waiting for the insects just as patiently as the zombifying fungus does. Fortunately, there are still enough active broods that periodical cicadas can usually be found somewhere in the United States almost every year. And the Brood X emergence in 2021 delivered a bumper crop of periodical cicadas—and *M. cicadina*-infected zombies—for scientists to observe and collect.

During the 2021 emergence, researchers also got a helping hand from citizen scientists who were on the lookout for Brood X cicadas, supplementing entomologists' efforts with day-to-day observations that they recorded throughout the springtime. Two apps—iNaturalist, a joint project of the National Geographic Society and the California Academy of Sciences; and Cicada Safari, created by the Center for IT Technology at Mount St. Joseph University—enabled anyone with a smartphone to record and upload photos and geolocation data about periodical cicada sightings. Cicada spotters across 20 states added tens of thousands of images per day at the peak of the cicada emergence, and by the time Brood X activity subsided, app users had uploaded 561,181 verified photos and reports of periodical cicadas, Gene Kritsky, a professor of biology at Mount St. Joseph University, wrote in 2021 in a summary of the Brood X emergence for *American Entomologist*.[13]

According to Kritsky, observers logged the first appearance of a zombie cicada—an abdomen-less adult with a visible yellow spore plug—on May 30, 2021. By June 9, the number of zombie sightings had increased significantly with more than four out of ten cicadas in some areas sporting visible spore plugs.

In New Jersey, Princeton was literally buzzing with cicadas, with tens of thousands of the insects crawling on sidewalks, swarming on lawns and in parks, and flitting among trees in such dense concentrations that local news outlets were calling Princeton the state's cicada capital. Periodical cicadas have blanketed Princeton since the 1800s, and many of the trees that were home to the nineteenth-century cicadas are still standing. Cicadas tend to stick close to the location where they hatch, which is probably why the cicadas are still so numerous in Princeton but are entirely absent in nearby towns that decades ago cut down and paved over most of their old-growth trees.

For some Princeton teenagers, the cicada emergence was a chance for a rare adventure: finding out what periodical cicadas taste like. Matthew Livingston and Mulin Huan, two students at Princeton High School, founded an insect-eating club in 2020, when they were both sophomores. Livingston launched the club as part of a research project investigating alternative protein sources; though the club's menus originally revolved around crickets, the 2021 Brood X cicadas offered students a chance to sample a new flavor of insect—fried whole or ground up into powder that could then be mixed into cookies.[14]

Princeton students weren't the only ones sampling Brood X cicadas. As emerging cicadas spread across the East Coast by the millions, recipes spread across the internet—spicy popcorn cicadas fried with cumin and cayenne pepper; tempura cicadas with sriracha aioli; crispy, sugar-roasted cicadas in salad; sauteed cicadas in tacos; cicada-crusted steak; even chocolate-dipped cicadas for dessert. As nymphs, cicadas somewhat resemble shrimp in both texture and flavor, and chefs such as New York's Jeff Yoon, who shares the joys of insect-based cuisine through his company Brooklyn Bugs, demonstrated recipes and held tastings on multiple news outlets during the Brood X months, recommending that Brood X nymphs be collected and cooked as early in their emergence as possible, so they would be at their plumpest and meatiest.

But what would happen if you ate a zombie cicada? *M. cicadina*'s zombifying mechanisms only affect periodical cicadas, so consuming the fungus wouldn't turn a person into a zombie—but could the amphetamines in the spore plug give you a buzz?

There are many nice things about being a science nerd and having science nerds as friends. For example, if you tell your science nerd friends that an emerging cicada brood will include zombie insects with psychoactive spore plugs and that you'll be out looking for them, these friends will promptly ask if you plan to lick the zombies to see if the fungus will affect you. Licking things for science isn't as weird as it sounds; geologists sometimes lick rocks to identify them. Field geophysicist, disaster researcher, and science writer Mika McKinnon described the practice on the social platform X (formerly Twitter) as "a real strategy" for scientifically identifying minerals, either by their taste or texture (McKinnon also included the wise caveat that "not every rock is safe to lick.")[15]

But licking living insects is a lot riskier than licking rocks, so tasting infected cicadas is probably not an advisable strategy. What's more, you'd probably need to lick quite a few cicadas to notice any effect, according to Kasson.

"With regards to psilocybin or cathinone, the amount we detected [in the *Fungal Ecology* study] was too small an amount for a human dose," Kasson says. For a human to be affected, "it would probably take dozens of cicadas," he added. And as there are other toxic chemicals in the spores, eating zombie cicadas would likely make a person sick before they felt any kind of pleasurable high.

Because infected cicadas aren't in much danger of being harvested by humans for recreational drug use, they're free to focus on more important matters: getting busy with as many healthy cicadas as they can, and as quickly as possible.

"The host emerges once every 17 years, so there's a very tight window to maximize dispersal," Kasson says. If you're a fungal puppetmaster manipulating the behavior of your host to focus on mating, it makes sense to chemically hypersexualize your hosts.

Zombification through sexual transmission is an effective if unusual strategy for *M. cicadina*, and it's an equally unconventional avenue in zombie movies and games. Contagion in a fictional zombie plague typically spreads through acts of violence, as zombies tend to bite first and flirt later (if at all), and uninfected people typically focus on fighting off zombie attacks or running for their lives rather than yearning for a zombie hookup. And unlike zombies' undead cousins—vampires—zombies are rarely portrayed as attractive enough to seduce a prospective partner into overlooking the fact that they're undead. Vampires and zombies alike are driven by an irresistible need to feed on humans, but vampires tend to be charismatic, physically perfect, exceptionally beautiful, and experts at seduction—think of Jacob Anderson as Louis de Pointe du Lac in the AMC television series *Interview with the Vampire*, ageless Catherine Deneuve and her alluring companion David Bowie in the film *The Hunger*, or Robert Pattinson's sparkly, moody bloodsucker in all the *Twilight* movies. Zombies, with their gaping wounds and rotting body parts, not to mention their insatiable hunger for human flesh and fresh organs, usually go right for the jugular, which tends to kill the mood—if it doesn't kill their victims first.

But luckily for *M. cicadina*, cicadas don't seem to mind getting it on with a zombie. In fact, as far as cicadas are concerned, an abdomen-less and spore-spewing mate is still pretty darn sexy. And horror filmmakers may finally be getting on board with that notion, with the recent rise of a genre dubbed "zom-rom-com"—zombie romantic comedies—which bring a new twist to zombie stories by exploring the likelihood of zombies as prospective partners for sexual and romantic escapades.

As for the amorous cicada zombie that I found that spring day in New Jersey, I only had about a minute or two to admire it before it took to the air, flew off across the field, and was lost to sight, probably high on amphetamines and almost certainly looking for love. Maybe it found a partner—or, more likely, several—and shared its spores with them before it expired, creating a fresh batch of zombified adults and ensuring that there would be a new crop of spores lying in wait to infect the next generation of Brood X and turn them into sex-crazed, butt-disintegrating zombies, too.

And perhaps, on a sunny spring day sometime in May 2038, I'll visit that Princeton field again and find them.

5

ATTACK OF THE OOZE

Zombifying Viruses

"It was a virus. An infection. You didn't need a doctor to tell you that."

—SELENA (NAOMIE HARRIS), *28 Days Later*

What does it mean to be alive?

That question frequently pops up during zombie stories, usually at some point around Act Two once the presence of zombies is well-established and characters are pausing to take a breather from whatever they were doing during their desperate battle to survive. In certain ways, zombies behave like they're alive—even the ones that are acknowledged to be reanimated corpses, like the zombies in the films *Night of the Living Dead* and *Shaun of the Dead*, and in *The Walking Dead* comics and TV series. Like living creatures, zombies are mobile, responsive to stimuli, and hungry—seemingly ready at all times to hunt and devour something (or someone). They're physically proficient and may be much faster, stronger, and more athletic than seems humanly possible. Sometimes those strong, swift, and acrobatic zombies can also be cunning and capable of planning and executing fairly complex strategies, such as the intelligent, resourceful, and deadly "Alphas" in director Zack Snyder's over-the-top, zombies-overrun-Las-Vegas heist movie, *Army of the Dead*.

In rare cases, such as the zombie medical examiner Olivia "Liv" Moore in the CW series *iZombie*, the zombified person is still herself—more or less. Liv (Rose McIver) is able to retain her intellect and personality, and demonstrates near-normal reasoning, thought and behavior—as long as she periodically ingests human brains. By eating the brains of homicide victims, Liv absorbs their dying memories and helps solve their murders, in addition to preventing herself from becoming a monster. Win-win!

Other types of zombies are not quite dead but are nonetheless irreversibly transformed. Their brains and nervous systems are entirely at the mercy of an infectious manipulator, like the mindless, ravening, and bloodthirsty high school students in the South Korean dramatic series *All of Us Are Dead*, who are under the sway of the so-called "Jonas Virus."

Regardless of which category zombies fall into, the general consensus is that they exist in a state that is somewhere between living and nonliving. And as it happens, there's something in the natural world that isn't a zombie but also inhabits that same murky area: viruses.

But hold on, you might say: Aren't viruses living organisms like other microbes, such as bacteria, protists, molds, and algae? The answer is more complicated than you may expect and has been vigorously debated for centuries.

Scientists first identified viruses in the late 1800s, describing them as infectious agents that were even smaller than bacteria (which had been identified about a century earlier). Viruses are also much smaller than the cells they infect. Take the rhinovirus, for example, which causes the common cold; about 500 million rhinovirus particles would easily fit on the head of a pin. But viruses can vary greatly in size. Most land in the range between 0.02 and 0.3 micrometers (roughly 0.0000009 to 0.000012 inches) in diameter, though there are some so-called giant viruses that have much bigger genomes and are the size of bacteria, boasting a diameter of a whopping 1 micrometer (0.00004 inches). Viruses live on every continent and in every ocean, and billions of them ride air currents worldwide—free-floating or hitchhiking on water vapor or soil particles—and then drift down to the ground. Hundreds of millions of these floating viruses shower each square meter of Earth's surface every day.[1]

Viruses come in four shapes: spherical, polyhedral, helical (coil-shaped), and complex, in which a polyhedral head perches atop a helical body, circled at the bottom by a ring of tail fibers resembling spiders' legs. A virus has a core genome that contains a single strand or a double strand of either DNA or RNA, tucked inside a protein shell called a capsid. Some viruses have an additional protective coating—a lipid membrane called an envelope—wrapped around the capsid. Viruses such as coronavirus, human immunodeficiency virus (HIV), and influenza are enveloped viruses. Viruses that lack an envelope—they're also known as "naked viruses"—include adenovirus, enterovirus, norovirus, and rhinovirus.

Whether a virus is enveloped or naked, it can only reproduce by invading a host cell. First, a virus attaches to a cell wall and is engulfed. It then releases its DNA or RNA inside the cell, where its genetic material replicates and forms new viral particles by hijacking the cell's energy-producing machinery, either in the nucleus (for DNA viruses) or in the cytoplasm (for RNA viruses). Newly formed viruses exit through budding—attaching to the cell membrane and then stealing a bit of it to wrap around themselves like a cloak—or through lysis, which ruptures the cell wall. Then off they go to invade and infect more cells, typically leaving a trail of dead or dying host cells in their wake.

For at least four billion years, viruses have populated the planet, and they exist wherever there is life. They respond to their environment and can evolve, as all living things do. But while viral reproduction is intertwined with living organisms, viruses' inability to make more viruses on their own challenges traditional definitions of life. They also are incapable of producing energy through metabolic processes—another hallmark of life—and are inert unless they are invading a cell or replicating inside it, using the cell's own metabolism as a battery.

At the time of their discovery centuries ago, viruses were regarded solely as agents of infection and disease; the word "virus" is derived from the Latin word for "poison" or "toxin." Indeed, viruses are responsible for many types of illness: COVID-19, polio, meningitis, measles, smallpox, the flu, herpes, hepatitis, and Ebola are just a few examples. But not all viruses are pathogenic—disease-causing—and many types of viruses are in fact intertwined with the human body, whether we like it or not. Scientists have learned over the past few decades through analysis of the human microbiome—the microbial communities that coexist inside our bodies—that each human individual hosts at least as many bacteria, viruses, and fungi as we have cells, with approximately 380 trillion viruses alone living inside us and on the surface of our skin. Some of these viruses are capable of causing disease, but many are neutral or may play a role in maintaining our health by bolstering the immune system or targeting harmful bacteria. As a species, our interactions with viruses over evolutionary time have even introduced bits of viral RNA into our genetic code—roughly 8 percent of the human genome is made up of fragments of retroviruses.[2] Viruses that infected our ancestors left behind these viral calling card remnants, which were then inherited by their hosts' descendants. Most of this genetic material is incomplete or highly mutated—or both—and doesn't interact with any of our functional genes anymore.

To date, virologists have described thousands of species of viruses and have discovered hundreds of thousands of new viruses that are awaiting classification. In fact, you're hosting a party in your digestive system right now with many of those unidentified viruses in attendance. In 2021, scientists from the United Kingdom and Europe identified more than 140,000 viral species living in the human gut, and about half of those species were unknown to science.[3] As many as 1.7 million viruses that are still unknown species are thought to infect mammals and birds; of those, between 631,000 and 827,000 could potentially make the evolutionary leap from wildlife into humans.[4] By some estimates, the number of viral species that are yet to be found worldwide may number in the trillions. As for the scope of individual viruses inhabiting our planet, the number is thought to be 10^{31}—10 followed by 31 zeroes—more stars than there are in the universe. A single teaspoon of seawater may hold up to one billion viral particles. And if all of Earth's viruses were assembled in a row, the line would extend for 100 million light years.[5]

Because viruses are commonly associated with infectious disease, they often turn up in horror games, movies, and TV shows as a driver of zombification. In reality, their transformative effects are generally somewhat less severe, but that's not to say that viruses are entirely innocent when it comes to host manipulation. Certain viruses that attack the central nervous system—such as viruses in the *Lyssavirus* genus, which cause the disease rabies—can affect the host's behavior. In some cases, viruses may leave behind a parting gift of lingering behavioral changes even if the infection is cured.

As it happens, there's one group of viruses in particular—baculoviruses—that's well-known not only for causing notable behavioral change but also for leading their hosts to an extremely gooey zombie-like demise. We should consider ourselves very, very lucky that the virus family Baculoviridae only affects invertebrates. Baculoviruses are found in more than 700 insect species, mostly in the orders Diptera (flies), Hymenoptera (ants, sawflies, bees, and wasps), and Lepidoptera (moths and butterflies), and in the crustacean order Decapoda (lobsters, shrimp, crayfish, and crabs). They are a group of enveloped viruses with genomes made of double-stranded DNA, and are rod-shaped; the name "baculovirus" is derived from the Latin word "baculum," which means "staff" or "stick." For more than a century, scientists have been studying the gruesome impacts of baculoviruses on caterpillars, especially those of leaf-munching agricultural and forest pests such as the small mottled willow moth (*Spodoptera exigua*) and nun moths (*Lymantria monacha*). But even though the larvae of these moths can dish out considerable damage to crops and forest habitats, they don't deserve the fate that accompanies infection by a baculovirus.

The virus first manipulates caterpillars into unnatural and hyperactive eating and climbing behaviors; however, its zombification story doesn't end there. Caterpillars that fall under the sway of the virus seek out the treetops near the canopy—or the highest point on a plant that they can find. Once they get there, they die, begin to dissolve, and then eventually dribble out of existence. They cling with their hind limbs and swing head-down from trunks, leaves, and branches high in the treetops during the infection's final stage. After they die, their weakened tissues rapidly decompose, and their bodies melt into caterpillar soup—an extraordinarily messy end that rivals anything dreamed up by special effects artists in the goriest zombie movies. Gooey former-caterpillar glop is loaded with viral particles, and each liquid caterpillar droplet that patters onto vegetation below can all too easily spread the baculovirus to new larval hosts.

In the summer of 2017 on a stretch of windswept moorland in the western United Kingdom, ecologist Chris Miller was conducting a butterfly survey for the Wildlife Trust for Lancashire, Manchester, and North Merseyside, when he discovered dead caterpillars dangling head-down from shrub branches. Further investigation in Winmarleigh Moss and across the West Pennine Moors revealed the grisly remains of other caterpillars that had suffered a similar fate. In some cases, only small scraps of their skin remained, after the virus led the caterpillars on an upward death march that ended with the insects' bodies liquefying into a sticky slurry. The sight of all those decayed caterpillar corpses was eerie and unnerving, Miller told the *Lancashire Telegraph* in July of that year.[6]

"It's like a zombie horror film," he said.

These caterpillars—larvae of the oak eggar moth (*Lasiocampa quercus*)—eat bilberry and heather plants and typically stay close to the ground, so it was "really unusual" to find them in elevated spots where they would be visible to predators such as birds, Miller told the *Telegraph*.

As early as 1891, atypical climbing behavior among infected nun moth caterpillars had already caught the attention of biologists, though at the time they didn't know what was causing the behavioral change. Nun moth caterpillars hatch from late April through early May in forests across Central Europe and in parts of Asia, and they immediately begin feeding on tender, young needles and floral buds on pine, spruce, larch, and fir trees. About once a week over the next two to two-and-a-half months, caterpillars level up in size by shedding their exoskeletons. They molt through five to seven progressively larger larval stages, known as instars, before forming a pupa and transforming into adult moths.

But during the nineteenth century, biologists began noticing and reporting that some nun moth caterpillars that were only partway through the instar process would start to lose interest in food and would move erratically. Within just a few days, the larvae would embark on a one-way trip up a tree trunk and toward the crown, climbing higher than they normally would. Scientists began referring to the syndrome as "Wipfelkrankheit," or tree-top disease.[7] The caterpillars' latter-stage symptoms were also described as "wilt disease," a reference to the limpness of their bodies as they hung from foliage head-down and dangling by their prolegs—the stubby limbs located toward the back of the abdomen—while their bodies dissolved into goo.

For a zombifying pathogen, piloting the caterpillars upwards before killing them has a couple of benefits. When a zombie

Figure 7 Caterpillars infected with baculovirus climb to the tops of trees, where they "melt" and drizzle the virus onto the foliage below. *Courtesy of David Cappaert*

caterpillar dies in an elevated spot and its body liquefies, the goop drizzles down onto other caterpillars, vegetation, and soil below, spreading the infection (see figure 7). Climbing to an elevated, highly visible position above the forest floor also exposes an infected caterpillar to predators, which may then help the pathogen by dispersing it across greater distances.

By the mid-1940s, scientists determined that this unusual wilt disease was caused by what are now known as baculoviruses. Insects become infected after eating something that has a baculovirus hitchhiker. The virus then travels into the midgut, where alkaline liquid dissolves its protective protein capsules and releases viral particles. As the virus begins replicating, it produces budded viruses that invade other body tissues. The virus then spreads into the brain and throughout the body, eventually liquefying tissues that then ooze out through ruptures in the insect's weakened exoskeleton, dispersing the virus across leafy surfaces that other insects are sure to encounter and eat—and become infected. And the infectious reach of just one sickened caterpillar is considerable; at the time of death, a single caterpillar corpse can contain more than 100 million clusters of viral particles known as occlusion bodies.[8]

Over decades, the more that researchers studied baculoviruses and the insects that the viruses infected, the more they recognized the deadly efficiency of the pathogens' mechanisms for manipulating and then annihilating their hosts. Because baculoviruses are widespread, highly infectious, and also very specific when targeting their favored hosts, they have naturally helped to curb population explosions in certain moth species, such as the spongy moth (*Lymantria dispar*, formerly known as the gypsy moth) in North America, a forest pest that was introduced as an invasive species in the late 1860s. According to a report published in 1990 in the *Annual Review of Entomology*, a type of baculovirus called nuclear polyhedrosis virus (NPV) can lead to the collapse of high-density populations of spongy moths.[9] In fact, baculoviruses do such a good job of regulating some types of insect outbreaks that by the 1970s, the World Health Organization and the United Nations' Food and Agriculture Organization recommended NPV and other related baculoviruses for use by humans as a form of insect pest control; unlike many conventional pesticides, the viruses are highly specific to their host insect species and have little environmental impact.

However, not all of baculoviruses' victims are considered pests. Silkworms—larvae of the moth *Bombyx mori*—have been cultivated in China for nearly 5,000 years for their silk-spinning prowess and are vulnerable to NPV infection. Disease from NPV has been documented in silkworms for centuries. It was sometimes referred to as jaundice, for the yellowish discoloration it caused in the silkworms' plump bodies, which are usually white tinged with pale green. But the disease is more commonly known as grasserie—from the French word *grasse*, which means "sticky"—as a reference to the goopy mess that NPV makes of infected silkworms' bodies, wrote USDA biologist Rosalind James and Zengzhi Li, a researcher at Anhui Agricultural University in Hefei, China. Infected silkworm larvae behave differently from uninfected caterpillars, "becoming easily agitated when disturbed," James and Li reported in the textbook *Insect Pathology*.[10]

The virus also actively directs the caterpillars' movements. Caterpillars are eating machines, spending most of their waking hours feeding their tiny faces. But as they near the end of the last growth cycle in their larval stage, they become more mobile and begin to wander. In healthy caterpillars that are ready to pupate—where they form a chrysalis and begin metamorphosing into an adult—wandering helps them find a safe spot to spin their protective pupae. But baculoviruses hijack that mechanism. By boosting the silkworms' attraction to light, the virus induces an infected caterpillar to instead wander upward, where it dies before metamorphosis can begin, scientists reported in *Proceedings of the National Academy of Sciences* (*PNAS*).[11]

For the *PNAS* study, researchers from the United States and Japan infected silkworm larvae with the silkworm baculovirus *Bombyx mori* NPV, or BmNPV (many baculoviruses are species-specific, so scientists frequently classify the viruses by naming them after the first insects that a virus was known to infect). About three days after infection, caterpillars showed the first signs of unusual locomotion. Their wandering behavior increased and then peaked after about four days, with the most wandering taking place around 12 to 24 hours before the caterpillars died—and NPV-infected caterpillars were more attracted to light than uninfected larvae. As the disease progressed, the caterpillars' skin became fragile and ruptured easily, spilling out the milky hemolymph inside. Silkworms at this stage left visible trails of virus-laden "blood" behind them as they crawled, which likely helps the virus spread to other caterpillars.

Next, the researchers needed to identify how the baculoviruses were manipulating their caterpillar hosts. The scientists generated 27 mutant strains of BmNPV, each with a different gene deactivated, and then introduced the viral mutants into caterpillars and observed the insects' behavior. Caterpillar locomotion was changed by all the mutant baculovirus strains—except one. Viruses lacking the *ptp* gene, which produces the protein tyrosine phosphatase, failed to induce light-activated wandering behavior in the infected caterpillars.

In 2011, another team of scientists from Pennsylvania State University and Harvard Medical School unlocked an additional genetic clue to the viruses' control of caterpillars' climbing behavior, this time in spongy moths infected with the *Lymantria dispar* multicapsid NPV (LdMNPV). Spongy moths are found across Africa, Asia, Europe, and North America; females lay egg masses in the fall that may contain a few hundred to more than a thousand eggs. When the caterpillars hatch in spring and early summer, they measure no more than 0.25 inches (0.6 centimeters) long. They spend about seven weeks as larvae, growing bigger by molting through five to six instars. By the time they reach the final instar and are ready to form a pupa and transform into adults, males typically measure 1.5 to 2 inches (3.8 to 5.1 centimeters) long and female caterpillars are 2.5 to 3 inches (6.4 to 7.6 centimeters) long; their bodies are covered with threadlike, barbed hairs, and long yellowish tufts sprout from their heads. The larval stage is when they do all their eating—and that's when the zombifying baculovirus takes hold.

Over the first few instars—each of which lasts about a week, depending on the environment and weather—the caterpillars stay near the treetop branches day and night, and feed voraciously on leaves. When they reach the fourth instar, they become more circumspect in their feeding, inching their way up toward leafy branches at dusk to eat through the night, and then descending and spending daytime hours lower down on the trunk of the tree, hiding from predators by hunkering in gaps between pieces of bark or in leaf litter. However, fourth instar caterpillars infected with LdMNPV don't descend their host trees during the day, and they don't molt. Rather, they linger on leaves out in the open and keep eating, gradually slowing down as the virus multiplies in their bodies and finally ends their lives.

Kelli Hoover, a professor of entomology and a researcher at The Pennsylvania State University's College of Agricultural Sciences, studies interactions between baculoviruses and insects, and how chemical signals produced by plants can shape those infections. In the late 2000s, as her research team was investigating that relationship by infecting spongy moth caterpillars with LdMNPV and then observing their behavior, they noticed something unusual.

"When we used a baculovirus that had a knockout of the *egt* gene—it had been mutated so it wasn't functional—we were seeing a difference in the behavior that occurs just prior to death," Hoover says.

That baculovirus gene's full name is quite a mouthful: ecdysteroid uridine 5′-diphosphate-glucosyltransferase, or *egt* for short. The *egt* gene produces a similarly named enzyme—EGT—which was known to deactivate a hormone in the larvae called 20-hydroxyecdysone. This hormone plays a critical role in the caterpillar's development, flooding the body when it's time for the larva to molt, and broadcasting that it's time to move on to the next instar. Caterpillars typically descend trees to molt, but if EGT blocks the release of that hormone, it removes important pages from the insect's biological instruction manual. Without the hormone signaling it to stop eating and head toward the ground to start molting, the caterpillar instead remains in the upper part of the tree and keeps on devouring leaves. Meanwhile, the virus continues to multiply until the caterpillar experiences a literal full-body meltdown.

The scientists suspected that infected caterpillars stopped descending to molt because EGT released by the virus was interfering with that hormonal signal. They set out to test that hypothesis, reporting their results in *Science*.[12]

Some caterpillars were infected with unaltered, EGT-producing baculovirus strains. The study authors found that infection in these larvae disabled their molting enzyme, and the caterpillars climbed as high as they could within their laboratory jars—and then died there. Another strain of LdMNPV with a disabled *egt* gene still killed the caterpillars that it infected—but those caterpillars died after descending to the bottoms of their jars. This told the scientists that the *egt* gene was responsible for repressing the caterpillars' natural inclination to climb downward, resulting in caterpillars that lingered in treetops until the virus killed them.

It's not yet clear how long ago LdMNPV acquired the means to control spongy moth climbing behavior. But this particular gene was stolen from the spongy moth genome, Hoover said.

"In fact, baculoviruses have several genes that they got from their hosts—just like we have viral DNA in our genome," she explains. This happens through a process known as horizontal gene transfer, which is a nonsexual transmission of genetic information between unrelated organisms.

"It's a lot more common than we realize," Hoover says.

Another key factor in the success of baculovirus' behavior manipulation is the presence of light. In 2017, scientists in the Netherlands and the United Kingdom found that timely exposure to light triggered the caterpillars' zombie-like climbing.[13] In prior studies from just a few years earlier, the researchers had demonstrated that light attraction, or phototaxis, was a motivator for infected caterpillars' unnatural climbing. The researchers then discovered that the timing and position of light exposure was important, too. They observed baculovirus-infected third-instar caterpillars of the small mottled willow moth, which they kept in glass jars lined with mesh for the insects to climb. The larvae would typically start climbing around 57 to 67 hours after they were first infected. But during experiments in which the larvae were exposed to light from different directions, caterpillars infected with *Spodoptera exigua* multiple nucleopolyhedrovirus (SeMNPV) only climbed upward in response to light shining from above them. If the light source was placed below the caterpillars, they stayed at lower positions on the mesh until they died.

What's more, the light exposure had to take place 43 to 50 hours after infection set in; otherwise the larvae wouldn't climb and would instead die "at low positions" in their jars, the study authors reported. Exposing caterpillars to light either before or after that critical seven-hour window also failed to trigger upward climbing. The scientists hypothesized that, clearly, the baculovirus was commandeering light perception pathways in the host's central nervous system in order to make the larvae climb—but it could only do so at a very specific time during the infection.

Except, how was the virus doing that? In 2022, another research team from China, the United States, and the Netherlands investigated the possibility of a link between the caterpillars' climbing and how the virus affected their perception of light. Their experiments identified gene pathways to explain how a baculovirus could hijack caterpillar vision, in larvae of the cotton bollworm moth (*Helicoverpa armigera*) that were infected with HearNPV (*Helicoverpa armigera* nucleopolyhedrovirus).

First, the researchers showed that infected larvae had stronger phototactic responses than uninfected caterpillars did—and the higher the light source, the higher they would climb. Research over the previous few years had identified genes linked to visual perception in *H. armigera* adults, "but the expression of these genes has not been studied in larvae," the scientists wrote in *Molecular Ecology*.[14]

The authors identified three genes that produced light-sensitive proteins that were critical for the caterpillars' vision system; they also learned that the genes expressed their proteins differently during HearNPV infection. Those genes were *HaBL*, for detecting short-wave light; *HaLW*, for detecting long-wave light; and *TRPL*, for helping cell membranes convert light into electrical signals, a process known as transduction. When the researchers disabled those three genes, using the CRISPR/Cas9 gene-editing system, the caterpillars' responses to light and subsequent climbing behavior were "significantly reduced," the scientists wrote.

"These results reveal that HearNPV alters the expression of specific genes to hijack host visual perception at fundamental levels—photoreception and phototransduction—in order to induce climbing behaviour in host larvae," the authors wrote. "Our present study not only confirms that baculoviruses induce host phototactic responses in the system of HearNPV and *H. armigera* larvae, but also provides valuable information to further understand the molecular mechanism of this behavioral change."

For baculoviruses, all these manipulative tricks work together to achieve one goal: produce a zombified host that behaves in abnormal ways to benefit the virus. In fact, "baculovirus-induced hyperactivity and tree-top disease are considered classical examples of parasite-mediated host behaviour," scientists at Wageningen University & Research in the Netherlands wrote in 2015.[15]

Tree-climbing and hyperactivity certainly make the caterpillars more visible to hungry birds, which may also benefit the virus. In North America, some of the bird species that prey on spongy moth caterpillars, such as the red-eyed vireo (*Vireo olivaceus*) and the black-capped chickadee (*Parus atricapillus*), swallow smaller larvae whole but take a slightly different approach for eating bigger, late-instar larvae—the birds tear them to pieces.[16] By ripping the caterpillars to bits to get to the tastiest inner morsels, birds that unknowingly target infected larvae may play a part in scattering the viruses more widely across the spongy moths' habitat.

Dying in an exposed place at a higher elevation can help the virus spread through another grisly path that may sound disturbingly familiar to fans of zombie movies: cannibalism. But unlike the films that have zombies devouring uninfected people, baculovirus-infected and zombified caterpillars are the ones that are being eaten.

Spodoptera exigua larvae are more commonly known as beet armyworms, and though the small willow moth species originated in Asia, these invasive agricultural pests are now commonly found throughout the world. Beet armyworms feed on at least 50 plant species, including crops such as alfalfa, corn, cotton, lettuce, onions, peas, potatoes, soybeans, tobacco, and tomatoes, to name just a few.

On occasion they also eat each other, should they happen to come across a fellow caterpillar that's already dead. This behavior, known as necrophagy, may happen more frequently when the recently deceased corpse fell victim to the baculovirus SeMNPV, according to scientists with the Instituto de Ecología A.C. in Veracruz, Mexico, and the Instituto de Agrobiotecnología in Pamplona, Spain.

During greenhouse experiments with SeMNPV-infected beet armyworms, the researchers noticed that healthy caterpillars were eating dead, infected larvae that were hanging from the upper reaches of the plants. While this gruesome meal may provide caterpillars with some essential nutrients, such as fatty acids and proteins, dining on dead larvae is also very risky, as it greatly increases the chance of SeMNPV transmission. So the scientists wondered if perhaps the virus was producing an alluring chemical signal that made the infected corpses smell or taste more attractive and delectable to healthy beet armyworms.

To test that hypothesis, the scientists set up a tempting buffet of fourth-instar caterpillar cadavers, offering healthy larvae a choice between beet armyworm corpses that were virus-free and corpses of larvae that died of SeMNPV. They checked healthy larvae's choices in two ways: by placing them in "Petri dish arenas"—Y-shaped acrylic tubes leading to two separate chambers, one with an uninfected corpse and one with an infected corpse—and by observing larvae as they foraged on sweet pepper plants in a greenhouse.

When released into the acrylic arenas, healthy larvae didn't exhibit a special preference for infected corpses. Rather, they selected (and ate) uninfected corpses about as often as they chose infected ones, suggesting that they were attracted to caterpillar remains "whether there was a virus present or not," the scientists reported in *PLOS ONE*.[17]

However, that wasn't the case in the pepper plant experiments. There, necrophagy happened more frequently on infected caterpillar remains, with 73 percent of the healthy larvae snacking on infected corpses. But those cadavers might be more attractive meals simply because of their location, rather than their extra-delicious flavor or odor, the study authors proposed. The SeMNPV baculovirus compels infected beet armyworms to climb to the tops of plants and then hang there until they die. When healthy caterpillars visit the tops of plants in search of young nitrogen-rich leaves, they are more likely to find a mealtime bonus there in the shape of a tasty corpse—which also happens to be brimming with a highly infectious pathogen. Once a beet armyworm takes its first bite of an infected corpse, its fate is sealed. In all the experiments, all caterpillars that fed on infected larvae sickened and died of SeMNPV. By compelling caterpillars to climb to their deaths, the virus increases the chances of the corpse being eaten by another caterpillar, and thereby ensures that the virus passes to a new host.

Do humans have anything to fear from baculoviruses? Our lives are already more intimately connected with them than you might think, through baculovirus-infected insect hosts that are commonly found on many of the crops that we grow.

Take the cabbage looper (*Trichoplusia ni*): the plump, bright-green larvae of this moth species get their name from the looping motion of their bodies as they crawl and from their seemingly insatiable appetite for cabbage and related crops such as cauliflower, broccoli, and brussels sprouts. Where there are cabbage loopers, there are cabbage looper–infecting baculoviruses, known as *Trichoplusia ni* NPV. Caterpillars that ingest the baculovirus decline over the next five to seven days, according to the late entomologist John Capinera, professor emeritus of integrated pest management at the University of Florida. As infection progresses, the caterpillar's green color fades to white, and its body becomes limp and bloated. "Death usually follows within hours following the limp condition, and caterpillars are often found hanging by their prolegs," Capinera wrote in a university outreach publication. After death, dark blotches spread across the caterpillar's body, and its skin weakens and tears easily. When the fragile skin ruptures, liquefied body tissues spatter down to the foliage below, where the droplets—and their viral payloads—will soon be eaten by other caterpillars.[18]

As it happens, caterpillars aren't the only ones unknowingly swallowing a payload of NPV in melted dribbles of cabbage loopers. Because the insects feed on crops, USDA scientists wondered if baculoviruses from melted cabbage loopers might linger on vegetables, traveling from fields to supermarkets to tables. In the 1970s, when researchers sampled cabbages from produce shelves in five supermarkets in the Washington, DC, area, they found ample evidence of baculoviruses in all the samples. Just one serving of cabbage—enough leaves to cover 16 square inches

(100 square centimeters)—contained up to 108 clusters of viral particles of *Trichoplusia ni* NPV. In one highly magnified image captured with a scanning electron microscope, a view of the underside of a cabbage leaf "shows the point in the leaf where an insect died and where the body contents have deposited masses of polyhedra [baculovirus' infectious protein crystals]," the scientists wrote in 1973.[19]

There are a number of reasons why this invisible viral payload on supermarket cabbage hasn't unleashed hordes of light-addled, tree-climbing, melting human zombies, but the main one is that baculoviruses don't replicate in human cells. We won't share the viscous fate of the zombified caterpillars because for that to be even remotely possible baculoviruses would be required to make a substantial evolutionary jump to mutate to the extent that they would find human cells to be hospitable hosts.

Even mutating to infect a new species is a big step for a virus. In viruses, certain protein "keys" enable them to unlock host cells, enter them, replicate there, and then easily transmit infection to new hosts. Such proteins might be a good match for the cells in one type of host but will fail to unlock the same access in a different type of host—for example, the virus might be able to enter a cell but not make copies of itself. That's where mutations can step in to lend a helping hand—to the virus, that is. Random mutations in viral populations may reshape their protein keys to better fit new cellular locks. If mutated viruses then encounter hosts with cells that respond to the new keys, that host species is vulnerable to infection.

As humans interact more and more with wildlife, viruses that historically infected only nonhuman animals have mutated and adapted to infect humans, in a process known as zoonotic spillover. Most infectious diseases in humans today have zoonotic origins; between 60 percent and 75 percent of known infectious diseases originated in nonhuman animals, researchers reported in 2021 in *Genetics and Molecular Biology*. Zoonotic diseases caused by viruses include acquired immunodeficiency syndrome (AIDS), rabies, zoonotic influenza (such as avian flu), West Nile virus, plague, monkeypox, severe acute respiratory syndrome (SARS), and COVID-19.[20]

All of the viruses that cause those diseases originated in vertebrates—mostly mammal species, such as bats and primates—before jumping to humans. Baculoviruses, which are presently best-known for infecting insects, evolved alongside insects about 300 million years ago, and their zombifying influence is currently reserved exclusively to the arthropod group. Recent genetic analysis shows that all major lineages of baculoviruses can be organized according to which arthropods they infect. The vast majority of baculoviruses are found in Lepidoptera hosts, so it's possible that their evolutionary relationship began when ancient baculoviruses infected ancestors of modern butterflies and moths, branching out from there to specialize in other insect orders and some crustaceans. Another explanation is that baculoviruses emerged around the same time as the first arthropods and then proliferated across certain insect orders, becoming especially successful in Lepidoptera.[21]

Scientists are still unraveling the evolutionary origins of this very special zombifying relationship. But as baculoviruses and insects coevolved over hundreds of millions of years, the viruses weren't the only ones building an arsenal of manipulating mechanisms and assembling toolkits full of zombifying techniques. Insects also explored that route, with some groups evolving weapons and behaviors capable of parasitizing and chemically puppeteering their fellow insects, as a novel (and usually gruesome) avenue for ensuring their own reproductive success. Over time, a number of insect species, rather than being zombified, became the zombifiers. As these zombie-makers embarked on that evolutionary journey, they developed uniquely diabolical methods for controlling their insect friends and neighbors—in ways that would make even the most seasoned naturalists shudder.

6

THE ORIGINAL CHEST-BURSTERS

Zombifying Wasps

"Game over, man!"

—PRIVATE FIRST CLASS WILLIAM L. HUDSON (BILL PAXTON), *Aliens*

One of the most memorable (and terrifying) scenes in sci-fi cinema history begins innocently enough. The crew of a commercial spaceship are seated around a circular table, dishing up a meal. They are relaxed, bantering and laughing, joking about the mediocre quality of the food that's available to them in space. Suddenly, one of the crew members begins to splutter, then to cough and choke. He collapses and convulses, falling on his back, writhing on the table and howling in agony as his alarmed friends hold him down and try to help.

And then the unthinkable happens—his chest bursts open, spewing blood everywhere. His friends barely have time to react before an eyeless, fist-size head on a wormlike torso punches through the man's rib cage from the inside. Gore and viscera spatter the stunned crew as they leap back from their comrade's feebly twitching body. The camera lingers on the blood-streaked creature as it pauses, its stalklike form poking from the ruin of the man's chest. Slowly, it pivots to survey the room, and then it utters a monstrous shriek, its mouth gaping to reveal rows of pointed teeth lining its jaws. Without warning, the ghastly thing abruptly bolts from the broken corpse and leaps to the floor, scuttles away and is gone, the crew left standing silent and frozen in a tableau of incomprehension and shock.

Written across all their faces is an expression mirrored in the faces of every person watching in movie theaters. The scene still transforms the features of anyone viewing the carnage for the first time, more than forty years later.

That expression says: "What in the *hell* did I just see?"

This scene was a pivotal moment in the 1979 movie *Alien*—it was the grisly outcome of a character's encounter with an unidentified extraterrestrial parasite. The chest-bursting creature remained nameless through the end of the film but was identified in the 1986 sequel, *Aliens*, as a "xenomorph"—the term comes from the Greek words *xeno*, meaning strange or foreign, and *morph*, which means shape or form. In the film, "xenomorph" isn't a scientific or common name for that particular species; rather, it refers to the creature's status as an unknown type of extraterrestrial. Like many real-world parasites, the xenomorph displayed multiple strange body plans over the course of its life cycle.

Its first form—a hatchling—was discovered by its first victim: a crew member named Kane (actor John Hurt), an executive officer on the cargo spacecraft *Nostromo*. In an earlier scene in the film, during a visit to an abandoned spaceship that had crash-landed on an unknown world, Kane found a clutch of hundreds of eggs. As he examined them, one of the eggs released an organism that looked somewhat like a large scorpion, with eight jointed legs and a long, flexible tail; it leaped at Kane's head, breached his helmet, and clung to his face. The creature remained firmly attached there while Kane lay unresponsive in the ship's sick bay; within a day it relinquished its grip on its host's face and was found dead nearby. Kane regained consciousness; though badly shaken, he initially appeared unharmed by the experience.

But unbeknown to Kane and the crew, during the creature's brief life stage as a scorpion-like "facehugger," it had inserted an embryo inside Kane's body.

This undetected parasite gestated rapidly. When it emerged, it looked nothing like the facehugger; its blunt, helmeted head tapered into a slender, armored body that ended in a serpentine tail, and the limbs were tiny nubs. After violently punching face-first through Kane's torso, the "chest-burster" stage of the xenomorph escaped into the ship and then continued to mature into its adult form, which had an elongated skull and pharyngeal jaws—a second extendable set of jaws inside its throat. This particular adult xenomorph was tailed, bipedal, and taller and more massive than a human. As the terrified *Nostromo* crew were hunted down one by one, they quickly learned that adult xenomorphs were dangerously efficient and deadly.

All of these life-cycle stages are uniquely terrifying, but the appearance of the chest-burster form was arguably the most horrific. In contrast to most of today's action blockbusters, the narrative pacing of *Alien* up to that point was deliberately slow, which made the carnage of the xenomorph's first appearance even more shocking. The film's gradual escalation of tension and dread effectively mimicked the patient, parasitic habits of the monster, said *Los Angeles Times* film critic Justin Chang.

"It takes its time because it means to infiltrate your nervous system," Chang explained. "Like our chest-bursting little friend, it wants to get inside you."[1]

*

While the horror of that scene is unforgettable, movie audiences can at least reassure themselves that the ghastly experience was purely an invention of science fiction.

Insects and spiders, on the other hand, aren't so lucky.

In their world, a chest-bursting creature's appearance—along with zombification that bends them to the parasite's will—originates in an encounter with a tiny female wasp armed with a long, slender stabbing structure on her rear end. This needle-like weapon, known as an ovipositor, dispenses numbing and development-arresting venomous brews, immune-system-suppressing viruses, and timed-release, body-destroying "bombs" in the form of eggs. Many species of these parasitic wasps reproduce by targeting larvae of Lepidoptera (the insect order that includes butterflies and moths), inserting their eggs deep inside caterpillars' bodies. Hatchling larvae feed exclusively on the living caterpillar's hemolymph, the blood-like fluid in invertebrates.

An infected caterpillar will never enter a chrysalis and become a butterfly. Rather, its metamorphosis will be arrested by a potent cocktail of metabolism-scrambling ingredients released by the wasp mother and by cells in her eggs, pausing the caterpillar's transformation so that it can continue to feed its parasitic hitchhikers, literally serving up its own body to them as breakfast, lunch, and dinner. Once the larvae are large enough, they chew their way to freedom and emerge through the caterpillar's skin. They spin cocoons and begin their own metamorphosis to become adult wasps, leaving their depleted hosts to die.

Wasps evolved this parasitic reproductive strategy many millions of years ago, and the fate that they visit upon their victims is so macabre that even the renowned scientist Charles Darwin found it profoundly disturbing. The nineteenth-century author of *On the Origin of Species* is widely credited as the father of the theory of evolution, and he was no stranger to odd animals and their peculiar habits. Darwin spent five years traveling the world onboard the HMS *Beagle*, as the ship's naturalist; from 1831 to 1836, he sailed around the South American coast and circumnavigated the globe. During that time, Darwin collected and documented thousands of samples of plant and animal species, pondering how evolution shaped their various forms and behaviors. He observed and recorded plenty of unusual and intriguing specimens: from Galápagos finches, to platypuses, to bizarre extinct animals like the giant mammal *Toxodon*—with its massive body, a face like a manatee's, and teeth resembling a rodent's, it was "perhaps one of the strangest animals ever discovered," Darwin wrote in 1845.[2]

Again and again throughout his travels, Darwin glimpsed first-hand the spectacular diversity in creatures around the world, and he recognized that such natural variation was vast, surprising, and essential to each species' survival. And yet Darwin still found it hard to stomach what parasitic wasp larvae in the Ichneumonidae family did to caterpillars.

In a letter that Darwin penned on May 22, 1860 to a friend and colleague—the American botanist Asa Gray—the great scientist professed that the gruesome habits of Ichneumonidae wasps raised serious doubts in his mind about whether or not the Christian god truly cared about his creations at all.

"I cannot persuade myself that a beneficent & omnipotent God would have designedly created the Ichneumonidæ with the express intention of their feeding within the living bodies of caterpillars," Darwin wrote. He further mused that the reproductive habits of ichneumonid wasps only deepened his growing conviction that the world was an utterly miserable place.[3]

Darwin's grim perspective may have been flavored by tragedy in his personal life; the death of his beloved daughter Annie in 1851 had already eroded his belief in a caring and benevolent deity. But Darwin also wasn't wrong in imagining a world that was shadowed by the mere presence of these wasps. There are certainly enough of them to literally darken the planet—by some modern estimates, over a million species of parasitic wasps are distributed globally, and parasitic wasps may comprise up to 20 percent of all insects on Earth.[4] They are known as parasitoids: a type of parasite that reproduces by laying eggs in the bodies of living hosts, where the larvae remain until they are ready to become adults, typically killing the host in the process. For parasitoid wasps, their preferred victims are usually insects or spiders. As the wasp larvae grow, they devour their hosts from the inside. But the host stays alive as they do so, usually until the developing wasp babies metamorphose into their mature forms. Most such wasps are tiny, with many species measuring less than 0.04 inches (1 millimeter) long as adults, though some grow to be around 3 inches (76 millimeters) or longer.

The wasp superfamily Ichneumonoidea contains two families of parasitoids—Braconidae and Ichneumonidae—that are also the two biggest families in Hymenoptera, the insect order that includes wasps, bees, ants, and sawflies. Scientists suspect that this massive superfamily may include as many as 100,000 species worldwide, though many of these species are yet to be discovered. In Ichneumonidae—the wasp family that attracted Darwin's attention and fueled his existential despair—there are more than 25,000 described species. The Ichneumonidae family name comes from the Latin word *ichneumon*; in ancient Greek it means "tracker." Today, ichneumonid wasps are also known as "Darwin wasps." The name was proposed in 2019 by an international team of entomologists to raise awareness about the insects, which, despite their notorious reputation, have been widely understudied since capturing Darwin's attention in the nineteenth century.

"We hope that the name catches on, and that Darwin wasps start buzzing more loudly across all disciplines of biology," the researchers wrote in *Entomological Communications*.[5] Whether Darwin would have been flattered or horrified by the gesture, we will never know.

Like the xenomorph in *Alien*, ichneumonid and braconid wasps mature through several different body forms in their life cycle; also like the fictional xenomorph, not every stage is parasitic. As adults, they are free-flying and nectar-drinking solitary insects that don't follow a queen as social wasps do, nor do they build hives or live communally. When females are ready to lay their eggs, they seek out an insect host from a species that is often specific to that particular parasitoid wasp species, and their target may be an egg, larva, pupa, or an adult insect. If the host is a caterpillar, being parasitized comes with a dose of hormones that suspend its metamorphosis. Further chemical manipulation of the victim's movement and other behaviors results in a zombified creature that lives only to fulfill its host's reproductive needs.

Female parasitoid wasps lay their eggs near, on, or in the bodies of the selected host. If they're caterpillar-seeking wasps, they may find their victims by following chemical distress signals called kairomones, which are volatiles—compounds that easily vaporize—released by some types of plants when hungry caterpillars come calling. For example, the braconid wasp *Cotesia marginiventris* can pinpoint potential hosts by tracking kairomone S.O.S. calls produced by caterpillar-damaged cabbage plants that are being chomped on by fall armyworms (*Spodoptera frugiperda*).[6] Think of it as screaming for help by emitting a certain type of smell.

Plants can even "scream" using different scent plumes depending on who's chewing on them, to attract parasitoids that specialize in that particular caterpillar species, scientists observed in *Nature* in 1998. *Cardiochiles nigriceps*, a species of braconid wasps, can only reproduce using tobacco budworm hosts (*Chloridea virescens*). The researchers observed that *C. nigriceps* wasps zeroed in on plants that were under attack by tobacco budworms, but the wasps were less responsive to distress calls from the same plants if they were being eaten by corn earworms (*Helicoverpa zea*). Chemical analysis of the plants' emissions using gas chromatography—a technique for separating and analyzing gaseous compounds—showed that plants emitted differing concentrations of volatiles in response to attacks from different caterpillar pests.[7]

"Without a specific signal from the plant, the parasitoid might otherwise spend all its time tracking false cues," said W. Joe Lewis, coauthor of the *Nature* study and an entomologist for the US Department of Agriculture (USDA) in Georgia. "It's like finding a needle in a haystack," Lewis told the USDA's *Agricultural Research*. "It has to find this one particular host in order to reproduce."[8]

Whether injected inside or attached to the host's exterior, when the wasp eggs hatch and the larvae emerge, the host transitions from being an incubator to being an all-you-can-eat buffet. A pupal stage soon follows for the wasp larvae. For some species that hatch inside a caterpillar, the youngsters chew tiny holes in their host's skin and squirm through before spinning their cocoons, which are tethered by silk threads to the caterpillar's body.

Such is the fate of large, plump caterpillars known as hornworms, when they are parasitized by the braconid wasp *Cotesia congregata*. Hornworms are so-named for the spiky structures that grow from their rear ends. These sizable horned larvae belong to the moth family Sphingidae; adults in this group are commonly known as hawk moths or sphinx moths, and are so big that they can easily be mistaken for hummingbirds. Two of the most widespread hornworms in North America are tomato hornworms (caterpillars of the five-spotted hawk moth *Manduca quinquemaculata*) and tobacco hornworms (larvae of the hawk moth *Manduca sexta*). The caterpillars of these two species feed on many of the same plants, which include tomatoes, potatoes, peppers, eggplants, and tobacco, and the caterpillars are similar in appearance. They grow to be around 4 inches (10 centimeters) long, and their voracious appetites make them notorious pests in gardens and crops across North America and Australia. Both hornworm species have bodies that are bright green with a horn at the back; but the tomato hornworm has a black horn and white

V-shaped markings, and the tobacco hornworm has a red horn and white stripes.

However, those identifying white marks can be hard to spot if *C. congregata* wasps are on board. A single parasitized hornworm may host many dozens of the wasps' wee, white, oblong cocoons, which cover its body like rows of very small seed pearls.

C. congregata wasps are found in North America, Central America, South America, and the West Indies, and they are known to parasitize more than a dozen species of moths and butterflies. Their tiny black bodies, which as adults measure just a few millimeters long, are dwarfed by their much-larger caterpillar targets. But when a female finds a promising caterpillar candidate to host her young, she takes advantage of her small size and swiftly makes her move, darting in to deposit her eggs inside its body with a needle-like ovipositor that's about 0.02 inches (0.5 millimeters) long. In just a few seconds, females can lay from a dozen to more than 100 eggs, each measuring no more than 0.006 inches (0.16 millimeters) in length.

And that's not the only gift that the mama wasp has hiding up her long and slender ovipositor. Along with the eggs, she delivers a critical ingredient for zombifying her victim: a virus that's produced in special cells inside her reproductive tract. Fluid surrounding the injected eggs is loaded with viral particles.[9] Researchers first discovered traces of these viruses in wasps in the late 1960s, but the journey that brought parasitoid wasps and viruses together began about 100 million years ago, when viruses in the family Polydnaviridae joined forces with the ancestors of some lineages of modern braconid and ichneumonid wasps.[10] Within Polydnaviridae are two genera of viruses: *Bracovirus*, which includes viruses that partner with some subfamilies of braconid wasps, and *Ichnovirus*, which includes viruses carried by some subfamilies of ichneumonid wasps. Polydnaviruses' genetic material, or provirus, is actually encoded into the wasps' genomes and is genetically distinct depending on which wasp species the virus inhabits. Proviruses are inherited by wasp young, and virions—infectious virus particles—replicate in female wasp pupae and in adult female wasps in calyx cells, which are located between the ovaries and oviducts.

The transfer of polydnaviruses, or PDVs, into caterpillar hosts serves a very important purpose. The virus expresses certain genes that disable cells in invertebrates called hemocytes, which typically envelope and destroy invasive organisms. This keeps the caterpillar's immune system from getting rid of the eggs and the hatchling larvae. PDVs also act as chemical roadblocks for certain hormones in the caterpillar that would normally flood the nervous system and kick-start the caterpillar's progression to the next instar—and its eventual metamorphosis.

The virus gets an additional boost from PDV-like proteins in the mother wasp's venom, which enhance the effects of the virus even as they slam the brakes on the caterpillar's development, optimizing the unfortunate insect as a host for wasp young.[11] But at this early stage of caterpillar zombification, the virus is doing most of the work; in fact, the wasps' entire reproductive strategy would probably collapse were it not for their virus partners.[12]

PDVs can't replicate inside caterpillars; at some point after partnering with parasitoid wasps, PDVs lost the genes that enabled them to copy themselves in cells of organisms other than their wasp hosts. The relationship they have with wasps is exclusive but also mutually beneficial; the viruses' own success depends on the wasp larvae thriving inside the caterpillar and becoming mature adults, enabling the virus to replicate. By domesticating viruses millions of years ago, the wasps ensured that their young would be kept safe and well fed, as they grew inside zombified caterpillars.

After a female *C. congregata* wasp injects her dozens of wee virus-coated eggs into a hornworm, the larvae hatch within just a few days. They are yellowish-white and grublike, measuring no more than 0.02 inches (0.5 millimeters) long. Though they can roam freely within the hornworm's body cavity, they typically stay close to the skin layer. Over the next eight days, they will grow to be around 0.09 inches (2.4 millimeters) long, and their bodies will develop more defined segments tipped with rows of curved bristles (these spikes disappear as the larvae get bigger). The second larval stage, or instar, lasts about another week. During this time, larvae nearly double in size, growing to measure 0.2 inches (5.0 millimeters) long.

And then they begin to chew through their host's skin (see figure 8).

Darwin may have been horrified by the sight of wasp young bursting out of caterpillars' backs, but nearly a hundred years later, an entomologist named Bentley Ball Fulton peered at what

Figure 8 White objects on the back of a tomato hornworm caterpillar (*Manduca quinquemaculata*) are cocoons spun by braconid wasp larvae. The cocoons are open, indicating that the wasps have emerged as adults and flown away, leaving their dying host behind. *Courtesy of G. J. Holmes*

was happening *inside* tobacco hornworms that bore developing wasp larvae, documenting and even illustrating every step of the gruesome process.

Fulton taught entomology at North Carolina State College, focusing on crop pests, from 1928 until he retired in 1954. Prior to that, he conducted entomology research at government-funded agricultural research stations in New York, Oregon, and Iowa. As a college professor in North Carolina, Fulton continued his entomology field research, and he described and sketched every developmental stage for *C. congregata* wasp larvae (at the time, the species was known as *Apanteles congregatus*), publishing his work in 1940 in *Annals of the Entomological Society of America*.[13]

Fulton scrutinized parasitized tobacco hornworms in North Carolina fields from August through October in 1936 and 1937, and he noted that in some habitats up to 82 percent of the hornworms were studded with numerous wasp cocoons. Female wasps "would land on a leaf near the worm, approach cautiously, and suddenly jump on the worm from the leaf," Fulton wrote. "The posterior part of the worm is the favorite point of attack and the worm immediately lashes its head about to drive the wasp away, so that the wasp is seldom on the worm [for] more than a few seconds."

But just a few moments was more than enough time for the wasps to deposit their egg payloads. When Fulton observed wasps and caterpillars in controlled laboratory experiments, he saw six wasps alight one after another on one of the caterpillars, each wasp touching down for no more than a second or two. Shortly after those multiple encounters, a dissection of the caterpillar revealed that the wasps had deposited a total of 122 eggs inside.

About two weeks later, when second-instar larvae were ready to exit a caterpillar, their heads were visible through the caterpillar's skin as they pressed up against its skin from inside and slowly nodded their noggins up and down, Fulton wrote. "The movement evidently serves to wear away the tissues of the host," he explained. The wasp larvae were also busily chewing as they nodded. "Under magnification the slender mandibles may be observed working in and out of the fold where they are concealed when at rest," Fulton continued. After wriggling through to the outside, the larvae molt one last time and then start spinning their cocoons, which remain attached at one end to the caterpillar's back, tethered by silk threads. The skin that they shed during their final molt shrinks and forms a tight plug that keeps the caterpillar from bleeding out, so it will stay alive as the cocoons take shape and until the wasp larvae exit them as adults.

The entire process—from egg-laying to back-bursting emergence—takes about 12 to 16 days, and the wasps emerge from their cocoons after another four to five days. There may be more than one brood of larvae per caterpillar, depending on how many female wasps deposited their eggs inside it, according to Fulton, who reported that the number of living larvae that eventually emerged from a single hornworm host varied widely, "from a few to over three hundred," he wrote. "Most of them come out of the host at the same time so that the back of the hornworm may be covered with cocoon spinning larvae."

This stage is an especially dangerous one for the developing wasps. They still need the caterpillar to stay alive—at least, until the wasps are ready to leave their cocoons, as being tethered to a dead host would expose them to scavengers and harmful bacteria. But caterpillars are also perpetually hungry eating machines that would almost certainly devour the attached pupae off their own backs if they had the chance.

That is, unless their wasp puppet masters were chemically manipulating the caterpillars to lose all interest in eating.

For the most part, the behavior of a parasitized caterpillar seems quite normal until about eight hours before the larvae begin nodding and nibbling their way out of its body and toward freedom. At that point, a caterpillar's movements become sluggish, and its normally bottomless appetite declines dramatically, according to Shelley Adamo, Charles Linn, and Nancy Beckage, from Dalhousie University, the New York State Agricultural Experimental Station, and the University of California, Riverside, respectively. They reported in the *Journal of Experimental Biology* that once the larvae began to poke their wee heads through the backs of infected tobacco hornworms, the caterpillars moved less and their interest in feeding dropped even more, and the study authors suspected that concentrations of a certain stress-related hormone produced in arthropod nerve cells might be involved.[14]

Octopamine is a neurotransmitter that helps relay neural messages linked to active movement, sending these signals to the arthropod nervous system. Adamo, the lead author of the study, wondered if the wasp parasites were suppressing octopamine production in tobacco hornworms when they emerged from their hosts, causing the caterpillars to move less and lose their appetites.

To find out, Adamo and her colleagues measured the amount of octopamine produced by infected and uninfected caterpillars. When octopamine levels were high in uninfected caterpillars, the insects became more active. In infected caterpillars around the time of wasp emergence, octopamine levels in their hemolymph were also elevated—about as much as in uninfected hornworms during their bouts of heightened activity. However, despite the abundance of octopamine, caterpillars with emerging wasp larvae moved less. They also spent much less time eating, took longer breaks between meals, and took fewer bites of their food.

There didn't seem to be any changes in their mouths or mandibles that might have prevented them from chewing and swallowing. Rather, "the tendency to feed, rather than the ability to eat, appeared to be most strongly affected by parasitism," the study authors reported.

The polydnavirus that the wasp mothers inject was already known to affect the caterpillars' immune response and arrest its metamorphosis. Could it be also responsible for behavior control at this late stage of wasp development? Previously, Beckage had coauthored a study with another UC Riverside entomologist—Mitchell S. Dushay—and found that injections of the virus alone weren't enough to deter caterpillars from eating. If anything, a hefty viral load seemed to stimulate caterpillar appetites, Dushay and Beckage wrote in the *Journal of Insect Physiology*.[15] While caterpillars in those experiments still didn't progress to the metamorphosis stage of their life cycle, they ate enough to pack on weight and often became "abnormally large" before dying, the scientists wrote. It was only after the wasp parasitoids emerged that the hornworms lost interest in eating.

One possibility that Adamo, Linn, and Beckage considered in their study was that the wasp larvae were somehow preventing octopamine from reaching the necessary neurotransmitters in a caterpillar's nervous system, thereby decoupling it from the messaging system that stimulated movement. Stress hormones accumulating in the caterpillar's hemolymph could play a part in decreasing its appetite.

Was that a good thing for the wasps? Adamo conducted further experiments on *M. sexta* caterpillars to see if their appetite loss directly benefited the wasp parasites, publishing another paper on the topic in the *Canadian Journal of Zoology*. Adamo found that nonparasitized caterpillars definitely had a taste for wasp cocoons and would devour up to 20 cocoons at a time if they found them lying on the ground. And in experiments when cocoons were glued to caterpillars that hadn't been parasitized by wasp larvae, the caterpillars eagerly ate the developing wasps right off their own backs, Adamo reported.[16]

But caterpillars that hosted the wasps from eggs to pupae were barely interested in eating at all once their hitchhikers emerged and left the sprouted cocoons alone. The caterpillars' reflexes and overall muscle tone were still strong, so they hadn't lost interest in food because they were at death's door. They simply stopped eating, Adamo wrote in the study. And that worked out just fine for the cocooned-encased parasitoids, who no longer needed to worry about becoming their host's dinner at a time when they were at their most exposed and vulnerable.

Clearly, cocooned wasps benefitted from undergoing their metamorphosis atop a zombified host that wasn't interested in eating them. And in 2016, Adamo and other researchers from Dalhousie, along with scientists from Acadia University in Nova Scotia and the State University of New York at Binghamton, discovered another important chemical clue suggesting how the wasps might be altering a host's neurochemistry to tweak their appetite and locomotion.

The scientists examined parasitized fifth-instar *M. sexta* caterpillars as *C. congregata* wasps emerged from the hosts' bodies, reporting their findings in the *Journal of Experimental Biology*.[17] First, the study authors confirmed that there were proteins that originated in wasps present in the caterpillars. They found that when the wasps exited the caterpillars, their hosts experienced a surge in the production of cytokines—proteins produced by immune system cells. A compound known as plasmatocyte-spreading peptide, or PSP, was especially abundant. And when the researchers injected nonparasitized caterpillars with PSP, the caterpillars ate less and gained less weight.

Prior findings about octopamine had already indicated that the stressed-out immune system in an infected caterpillar was working overtime when the wasp larvae burst through its skin. The new data about PSP told the scientists that this immune activity was even more complex and intense than once thought. Elevated levels of both PSP and octopamine suggested that the wasps stimulated a "massive immune response" in their hosts when they emerged, which then suppressed the caterpillars' impulse to feed. In fact, when compared with other types of immune challenges in caterpillars, "the exiting wasps seemed to induce an unusually large response," the authors wrote. It would be expected that a caterpillar's immune system would go to work as dozens of parasites punched their way through the host's back. But the scale of the immune response, according to the scientists' findings, hinted that the wasps might be augmenting this immune reaction, manipulating and magnifying it for their own advantage.

By this point, a parasitized caterpillar—slow-moving, barely eating, its back scarred and studded with cocoons—is probably more than ready for the sweet release of death. But metamorphosing wasps have one more job for their zombie host. Exposed cocoons are a tempting treat for other predators as well as caterpillars; they also attract insect parasites that prey on the wasps. As the final-instar wasp larvae slumber inside their pupae and their bodies transform into adults, their host takes on a new role. It becomes a dedicated zombie bodyguard that aggressively protects the parasitoids against attacks; the host caterpillar will actively defend its zombifiers, even though the parasites are no longer housed inside its body.

Entomologists Jacques Brodeur and Louise Vet, researchers at Wageningen University & Research in the Netherlands, published the first observations of this guardian behavior in 1994 in *Animal Behavior*, after watching caterpillars of cabbage butterflies (*Pieris brassicae*) as they defended cocoons of the braconid wasp *Cotesia glamorata.*[18] As adults, the wasps' tiny black bodies measure about 0.3 inches (7 millimeters) long. They lay about 20 to 60 eggs per caterpillar, and the larvae emerge and begin spinning their cocoons about 15 to 20 days later.

After *C. glamorata* wasps chew their way out of their hosts, they often build their cocoons nearby, attaching them to leaves. You might think that a host caterpillar, once relieved of its parasitic burden, would want nothing more than to surreptitiously creep away from those cocoons as quickly as possible. But instead, the caterpillars remained close by, coiling their bodies protectively atop the encased wasps and even spinning silk webs over the cocoons to add an extra layer of protection—an effort that took several hours, Brodeur and Vet reported.

The guardian caterpillars still wouldn't feed themselves, nor did their own metamorphosis resume. But they swiftly sprang into action if the cocoons were threatened. When the scientists tapped the caterpillars with the bristles of a tiny paintbrush (simulating a predator attack), the caterpillars displayed "a variety of aggressive reactions," such as flexing their legs, gnashing their mandibles, regurgitating droplets of red fluid, biting the paintbrush hairs, and "thrashing their head and thorax back and forth," the researchers wrote. These defenses became feebler after the adult wasps freed themselves from the cocoons and flew away, and yet the caterpillars still did not desert their posts. They eventually died near the pile of abandoned pupae, atop the remains of their silken shields.

Brodeur and Vet concluded that the parasites could seemingly still manipulate the caterpillars' active behaviors without the benefit of physical contact with their former hosts, which hadn't been observed before in parasitic insect interactions. The wasps appeared to be usurping defense mechanisms and strategies that a caterpillar would normally use to defend itself against attacks, weaponizing the caterpillar to keep themselves safe.

How exactly were they doing that?

One possible explanation emerged more than a decade later, proposed by another team of entomologists from the Netherlands and Brazil. They discovered that not all of the pupating wasp larvae abandon the zombified caterpillars; some of them stay behind after their siblings break free. Perhaps, the scientists suggested, by giving up the chance to spin a cocoon of their own, these larvae bought a few more days of safety for their brothers and sisters, compelling the caterpillar to guard the pupae by manipulating the insect from the inside.

For their study, published in *PLOS ONE*, the researchers examined caterpillars of the Brazilian geometrid moth (*Thyrinteina leucocerae*) that were parasitized by braconid wasps in the genus *Glyptapanteles*, a cousin of *Cotesia* wasps in the subfamily Microgastrinae (the group includes nearly 3,000 described species—though there may be tens of thousands of species that are yet to be discovered—and all of the known microgastrine wasp species parasitize caterpillars).[19] *Glyptapanteles* wasps, of which there are more than 120 species described worldwide, lay up to 80 eggs inside a caterpillar. When the larvae emerge, they spin their cocoons next to the caterpillar, attaching them to a nearby leaf, stem, or twig rather than to their former host's back.

First, the scientists wanted to confirm that the caterpillars' so-called bodyguard actions had something to do with the presence of the parasites. Experiments showed that after wasp larvae burst out of a caterpillar and prepared to pupate, their host stopped feeding and hunkered down close to the pupae. When predators came calling, the caterpillar turned into a headbanger to rival the most dedicated heavy-metal fan, knocking interlopers away from the cocoons by violently swinging its head. It then died without ever reaching adulthood as a moth.

None of these behaviors was demonstrated by nonparasitized caterpillars, the study authors reported. Of the 19 parasitized hosts that the scientists observed, 17 lashed out at predatory insect interlopers, such as several ant species, shield bugs (*Supputius cincticeps*), and other parasitoid wasps, warning them off "with repeated violent head-swings." Only one out of 20 nonparasitized caterpillars responded this way when a predator approached nearby cocoons. The rest of those parasite-free caterpillars barely responded to the threat to the cocoons at all—even when the would-be attacker strolled on top of a caterpillar.

Did the bodyguard caterpillars' aggressive head-swinging moves protect their cocoons? Field observations and lab experiments revealed that defended cocoons were safer when under the watchful eye of an attentive caterpillar guardian. "In more than half of the encounters of a predator with a parasitized caterpillar, the repeated head-swings caused the predators either to give up and leave the twig or to be knocked off," while predators that approached pupae with nonchalant caterpillars nearby were mostly undeterred from attacking their helpless prey, the scientists wrote. After the cocooned wasps emerged and flew away, the bodyguard caterpillars died anyway, so they didn't gain any benefit for themselves by fending off attackers. The purpose of this guarding behavior must therefore be to benefit the parasitoids that induce it, the researchers said.

As for how the wasps were compelling the caterpillars to be their feisty defenders, the former hosts' continued zombification may have been an inside job. When the study authors dissected one of the guardian caterpillars, from which dozens of living wasp larvae had exited just a few days earlier, the scientists found that there were one or two live larvae still lurking inside. The researchers hypothesized that these larvae were responsible for manipulating the caterpillar's bodyguard behavior; doing so meant that the left-behind larvae would perish when the caterpillar died and never spin their own cocoons, but their sacrifice promised their cocooned kin a better chance of survival. However, the study authors cautioned, further research would be necessary in order to pinpoint the molecular pathways that enabled the sacrificial larvae to control their bodyguards and compel them to defend their siblings.

*

Centuries ago, Charles Darwin lamented the grim outcome of ichneumonid wasps' reproductive habits. Scientists who study such wasps today typically have a somewhat less depressingly existential relationship with their subjects. Many encounter the wasps for the first time while rearing caterpillars and observing the tiny parasitoids as they pop out of their hosts, an experience that eventually leads many of those former caterpillar aficionados to stray from Lepidoptera, says entomologist Gavin Broad, a wasp taxonomist and principal curator in charge of insect collections at the Natural History Museum in London. Broad's own discovery of parasitoid wasps followed that path; he discovered the wasps while raising moths as an undergraduate student at university.

"Sometimes, instead of getting a moth when you rear the caterpillar, you get a nice little wasp that comes out instead. That's what converted me to the dark side," Broad says. "For some of us, it's much more interesting when you get the wasp out than the moth." He quickly realized that very little was known about ichneumonid wasps, and the scope of the group would provide fertile ground for investigation; he has since coauthored dozens of studies on parasitoid wasps, examining their taxonomy, diversity, evolution, and habits.

"When I first started studying them, what I wanted to do was just work on the true diversity of them and describe hundreds of new species in the tropics—which I still want to do," he says. But along the way, and with the estimation of many thousands of undescribed wasp species worldwide, Broad's focus shifted toward how the wasps partition up different ecosystems and how they determine what is a potential host and what isn't. "To test various theories about how speciation occurs in these wasps—that's what I'm really interested in," he says.

One standout example is sabre wasps (*Rhyssa persuasoria*); with bodies measuring up to 1.6 inches (40 millimeters) long, they are Britain's biggest ichneumon wasp species, and one of the largest in Europe. Females have ovipositors as long as their bodies, and they use these formidable structures to parasitize wood-boring larvae of insects such as horntail wasps (*Uroceris gigas*) and longhorn beetles (*Spondylis buprestoides*). Sabre wasps use their sense of smell to locate the hidden, burrowing larvae, then punch through wood with metal-enriched ovipositors to paralyze the larvae and lay their eggs inside them.

"That's a lovely example of how the inaccessibility of their hosts has driven some really extreme adaptations," Broad says.

Broad is also a coinvestigator for the Darwin Tree of Life project (DToL). The program, named in honor of Darwin's scientific legacy, continues his work of exploring and cataloging the vast diversity of life on Earth. The goal of the project is to sequence, assemble, and publish the genomes of more than 70,000 species—animals, plants, fungi, and protists—in Britain and Ireland. A substantial number of those species are wasps; there are about 6,500 species of parasitoid wasps in the United Kingdom, and 4,000 of those are from the wasp superfamily Ichneumonidea. In 2022, Broad and a team of researchers with the Darwin Tree of Life Consortium sequenced the project's first parasitoid wasp genome, for the species *Ichneumon xanthorius*, which is found in Europe, North Africa, the Middle East and Central Asia. This is also the first-ever reference genome for the Ichneumonidae family; the data is freely availablc in a public database for any researcher to access and use, the study authors wrote in the journal *Wellcome Open Research*.[20]

Darwin wasps may be ruthless zombifiers, but they're also an evolutionary success story; unlocking their genetic secrets could help scientists to better understand the adaptations that fuel the group's success and expose how exactly the wasps accomplish their host manipulation. One possible avenue for future investigations could include wasp olfactory genes, and how gene families are involved in detecting potential hosts, Broad says. And while caterpillars may be the wasps' most easily recognized hosts, those zombifying adaptations help Ichneumonidae wasps control a range of zombie victims. Other invertebrates—insects and arachnids alike—are susceptible to the wasps' mind-manipulating mechanisms, and may also find themselves coerced into serving and protecting developing wasp young. However, some of those hosts have evolved a few tricks of their own, helping them to fend off would-be zombifiers and, sometimes, even achieve the holy grail of horror movie zombie outbreaks: recovery from zombification.

7

'TIL DEATH DO US PART

Zombie Spiders, Roaches, and Ladybugs

"I will not turn into one of those things. Come on, make this easy for me."

—TESS (ANNIE WERSCHING), *The Last of Us* (video game)

Perhaps you've heard the story of the fictional spider named Charlotte who transformed her webs into a series of positive-affirmation billboards to save the life of a pig named Wilbur that she befriended. Despite all of Charlotte's supportive cheerleading, Wilbur was a fairly run-of-the-mill pig—runty, friendly, and cute, but with no discernible special abilities. Charlotte, however, was a true superstar. Night after night, she scripted praise for Wilbur in silk strands that she pulled from her body (a metabolically costly feat for a tiny spider), and her artistry in crafting webs that read "SOME PIG," "TERRIFIC," and "RADIANT" convinced Wilbur's owners to spare him from a farm pig's usual fate: the chopping block.

Charlotte introduced herself to Wilbur as "Charlotte A. Cavatica," which suggests she was probably a barn spider (*Araneus cavaticus*), an orb-weaver species that's common in North America. Orb weavers are in the spider families Araneidae and Tetragnathidae (long-jawed orb weavers), and there are more than 4,000 species of orb weavers worldwide (all spiders are grouped into 100 families, representing about 45,000 known species).

While orb weavers vary widely in color, size, and overall appearance, all are known for their wheel-shaped webs, with spokes that radiate from the center and that are joined by lines that cross the spokes in circles, forming a spiraling grid. Flying insects that blunder into an orb weaver's web become trapped in the spiral's sticky strands, while strong, nonsticky radius, frame, and anchor threads stabilize the web's structure and keep it from falling apart as the trapped insect struggles. It's a good system for prey capture, and orb weavers have been using it for about 136 million years.

To build their webs, spiders spin strands out of proteins that they produce in special glands as a liquid. They extrude the liquid from organs called spinnerets, located on the backs of their abdomens, and then use their legs to yank and pull the secretions into silk threads. Different glands produce a variety of sticky and nonsticky silks that are used for a range of tasks, including swaddling prey, ballooning, and making egg sacs. Most spider species have a cluster of silk organs; orb weavers typically have five.

Web designs can vary somewhat between orb-weaver species, but in general they follow the same basic construction pattern, says Samuel Zschokke, an arachnologist at the University of Basel in Switzerland. Zschokke's digital visualizations of web-making animate the process step by step; first, the spider builds a Y-shape of anchor threads, which connect in the middle. Next comes a frame around the perimeter, then more threads that radiate from the web's center to the outer frame, like the spokes of a bicycle wheel. The spider then attaches threads in a spiral pattern along the spokes, known as radii, starting at the center of the web and working outward. This is called the auxiliary spiral and is made of nonsticky silk. Finally, the spider retraces its steps from the outer frame back to the center, removing the placeholder spiral along the way and replacing it with sticky strands.

Even when orb weavers were sent into space by NASA for experiments onboard the International Space Station (ISS), they were unfazed by life in microgravity and continued building their trademark webs. Two "spidernauts"—golden orb spiders (*Nephila clavipes*) nicknamed Gladys and Esmeralda by astronaut Cady Coleman—visited the ISS in 2011. During their time in space, they produced webs that were more circular but otherwise much the same as their usual webs.[1]

However, back on Earth there are wasps that can make orb weavers forget everything they know about normal, prey-catching web production. Parasitoid wasp larvae overwrite orb weavers' internal web-making guidelines, compelling the spiders to instead build nonspiraling durable structures that cradle and protect pupating wasp larvae. The larvae kill and eat the spider once its work is complete, then retreat into the safety of their web-supported cocoons to become adults and continue the zombification cycle.

*

Ichneumonid wasps in Polysphincta—a group of genera in the subfamily Pimplinae—are known for parasitizing spiders and manipulating their behavior, taking advantage of the arachnids' web-spinning prowess and incorporating it into zombie guardian duties. There are 25 genera and more than 200 species of polysphinctine wasps distributed from Mexico to the southern part of Brazil. Some types of these wasps target arachnids' egg sacs, while others take aim at adult spiders. Female wasps that attack adult spiders typically lay a single egg per host, attaching it to the spider's abdomen. A wasp will set the zombifying process in motion by first stabbing a spider in the head (sometimes right in the mouth) with her ovipositor and paralyzing it with venom, to lay her egg. After the larva hatches, it piggybacks on its host's back until it's ready to pupate. That's when the larvae's zombification of its host becomes apparent, as the spider soon begins spinning a web that looks nothing at all like its usual prey-catching structure.

More than two decades ago, entomologist William G. Eberhard, then a researcher with the Smithsonian Tropical Research Institute at the Universidad de Costa Rica (now staff scientist emeritus), published the first detailed observations of spiderweb manipulation associated with wasp zombifiers. He wrote about the spider *Leucauge argyra*—a type of orb weaver, classified at the time as *Plesiometa argyra*—which was parasitized by the polysphinctine wasp *Hymenoepimecis argyraphaga*. At the time, there were few mentions in scientific literature of parasitized spiders' manipulated webs, and next to nothing was known about the behavior, Eberhard says. He'd previously noticed strange-looking webs during field work but hadn't investigated them too closely; eventually, he realized that the spiders that made these webs were carrying parasitoid larvae on their backs. Their "crummy webs," he says, were at least consistent in their crumminess. But while they may have been crummy by prey-catching web standards, they were very, very well designed to do something else: hold the cocoon of a mind-controlling parasite.

Eberhard's first paper on spiders' manipulated webs, published in 2000 in the *Journal of Hymenoptera Research*, documented interactions between wasps and their prey in Costa Rica, in plantation undergrowth and in the lab, describing 14 parasitizing attacks from start to finish. He reported that "there was always a brief struggle" at the start, but once a female wasp successfully stabbed a spider—delivering repeated venomous jabs—paralysis quickly set in and lasted for about five to ten minutes. The wasp used some of that time to probe the immobilized spider with her abdomen, seeking and sabotaging any eggs or larvae that had previously been deposited there by other female wasps. In one instance, a wasp dislodged a sizable larva from the spider's body and flicked it to the ground with her ovipositor, then proceeded to lay an egg of her own in its place.

Female *H. argyraphaga* wasps attached their eggs securely to a spider's abdomen with application of a special glue. When the larva hatched, about two to three days later, it remained partially inside the egg, using it like a saddle to ride its spider host. As the youngster grew through at least two instar stages (the exact total number is unknown, though it may be as many as five), it gnawed pinprick holes in the spider's abdomen and drank its hemolymph—never so much that it would kill the spider, as it needed to keep its host alive. When the larva got too big for its egg saddle, it constructed a new one, crafting it from accumulations of dried spider hemolymph and its own shed skin.

The spider seemed oblivious to the external hitchhiker, or ectoparasite; it behaved normally and spun its usual concentric radial, insect-catching webs, even as the larva repeatedly punctured its host's abdomen and sipped its body fluids. However, once the wasp larva reached its final instar and was ready to build its cocoon (about one to two weeks after hatching), the spider's behavior underwent a dramatic change. Though the spider didn't know it yet, this was its last day of life—and it was going to be eventful. By now, the larva has grown to be about one-quarter to one-third the volume of the spider's abdomen, and as it crouches atop its host, it compels the spider to begin spinning one final silk masterpiece. But this web won't be used for nabbing flies. Instead, the parasitized spider abandons its usual web pattern and spins a more simplified web. Cocoon webs, as Eberhard named them, have one purpose: to support and protect the pupa of its zombifier, which encases itself in a cocoon that dangles from the center of the modified web, suspended by silk threads.[2]

When Eberhard began collecting data on the modified webs, he was already very familiar with *L. argyra*'s normal web-building behavior, and one aspect of cocoon web construction quickly caught his attention. When a nonparasitized spider builds a radius, it typically runs out to the edge of its web and lays a line behind it. Once the spider reaches the web's edge, it breaks the line that it just made and retraces its steps back to the central hub, reeling up that strand and replacing it with a fresh one. Parasitized spiders building a cocoon web didn't break that first line, so on the way back to the hub, they were doubling the radius thread, Eberhard says. Somehow, the wasp larva's presence was inhibiting a behavior that is foundational for normal orb construction, and the manipulation was happening "in what looks like a really selective way."

In the cocoon webs he observed, a number of cables—reinforced radii made of multiple silk lines—emerged out of a central hub. Each cable had several branching ends that were all anchored to the surrounding substrate. An orb weaver's web is typically woven to be open in the center. But in the cocoon web's middle was a substantial tangle of silk threads, forming a dense and sturdy mat. In these modified webs, there were fewer radii and none of the orb weaver's trademark concentric rings connecting the radial lines. They also lacked a normal web's sticky spirals for trapping prey.

"These differences between cocoon webs and normal orbs all make the cocoon web a stronger, more durable support for the wasp's cocoon," Eberhard wrote in *Nature*.[3]

Orb-weaving spiders typically follow a very specific set of step-by-step instructions in order to build their daily webs. In zombified *L. argyra* orb weavers, the wasp larvae were blocking the spiders' access to most of the instruction manual, leaving their hosts with just a few pages from their web-building playbook. Zombie spiders had only one web-spinning option: cycle through a single "subprogram" of web construction over and over again, repeating one part of the process and excluding all others. Wasp larvae were likely manipulating their web-spinning hosts through chemical means rather than physical contact, "as the spider continues to build the cocoon web even when the larva is removed shortly before construction would normally start," Eberhard reported.

And the effects of that chemistry were not only fast acting but also long-lasting. If Eberhard removed a larva from a spider that had already started building a cocoon web, the spider doggedly persisted in its task despite the zombifier's absence. It was only after a few days had passed that a spider might slowly return to spinning its usual spiral webs.

While many parasites can manipulate behavior in a zombified host, most of those parasites—particularly insects—cause behavioral changes that are relatively simple, "such as movement from one habitat to another, eating more or less, or sleeping," Eberhard wrote. "*Hymenoepimecis*'s manipulation of its spider host is probably the most finely directed alteration of behaviour ever attributed to an insect parasitoid."

Polysphinctine wasps can even coerce social spiders into additional unusual behavior: leaving their friends and families to spin a cocoon web for their zombifying masters. There are about 25 species of social spiders across nine spider families worldwide, and such spiders live together in large groups all their lives. They rarely stray far from their colonies, which can contain tens of thousands of spiders. There, they communally care for young; build, fix, and defend their webs; and even work together in packs to hunt and subdue insect prey. However, researchers at the University of British Columbia in Canada observed something odd in *Anelosimus eximius*, a species of social spider in the Ecuadorian Amazon. An *A. eximius* spider that was parasitized by a larva from the *Zatypota* wasp genus would abandon its colony to spin a solitary, three-dimensional cocoon web between leaves.

For such tiny spiders, which measure no more than 0.2 inches (6 millimeters) long, voluntarily leaving the colony and striking out on their own is high-risk and extremely rare. And the cocoon webs made by the wandering spiders were unlike any that a social spider might build if it happened to be separated from its colony by accident. Texturally, the webs resembled the sheet webs that are spun collectively to form the foundations of a colony's immense web structure. But unlike those flat sheet webs, the webs that the roving, zombified spiders built were capsule-shaped, "entirely covered and enclosed with thick webbing; and had no apparent opening," the Canadian biologists reported in *Ecological Entomology*.[4]

More than anything, these webs resembled cocoon webs that Eberhard previously described in *Psyche: A Journal of Entomology*. Those webs were constructed by individuals from the solitary spider species *Anelosimus studiosus* and *A. octavius*, and both were zombified by *Zatypota* wasps in Costa Rica's San José Province. Their cocoon webs included flattened sheets of silk that wrapped above and below a central clump of densely woven threads.[5] But the *Zatypota* wasps in Costa Rica didn't compel the solitary spiders to wander before their manipulated web-building, as the social spiders in the Ecuadorian Amazon did.

Perhaps, the Canadian scientists considered, *Zatypota* wasps manipulated their social spider hosts by starving them; lack of food has been known to cause spiders to leave their colony. Another possibility was that the wasps were chemically unlocking long-buried ancestral genetic programming in the spiders' DNA, linked to dispersal—leaving a place of birth and traveling to another location to mate and raise young. The behavior is common across spiders, most of which disperse soon after hatching, but has been largely abandoned in social species.

The precise mechanisms of this behavioral change are tricky to unravel because overriding social spiders' natural inclination to stay with their colonies is a complex level of manipulation, said Philippe Fernandez-Fournier, lead author of the *Ecological Entomology* study. Fernandez-Fournier conducted the research while in the master's program at the University of British Columbia (UBC).

"This behaviour modification is so hardcore," added study co-author Samantha Straus, a researcher in UBC's Department of

Zoology. "The wasp completely hijacks the spider's behaviour and brain and makes it do something it would never do," Straus said.[6]

Whether a zombified spider is social or solitary, when its web-building work is done, so is its usefulness to the wasp—except as the larva's last meal before it enters the pupal stage. Shortly after an *A. studiosus* or *A. octavius* spider's cocoon web was complete, "the wasp larva then killed the spider, sucked its carcass dry and dropped it," Eberhard reported in *Psyche*. Fortified by its delicious spider dinner, the larva promptly began building its pupal cocoon, securing it to the underside of the densely woven central portion of the dead spider's final web. A larva would first spin a suspension line from multiple threads, then encase itself in a silk capsule from the bottom up. Transformation into an adult wasp took about 10 to 12 days, so a cocoon web had to be durable enough to hold together for that entire time, until the larva's metamorphosis was completed.

By the 2010s, scientists were hot on the trail of the wasps' secret weapon for changing spider behavior. Researchers from Japan observed the polysphinctine wasp *Reclinervellus nielseni* as it turned its host spider—the orb-weaver *Cyclosa argenteoalba*—into "a drugged navvy," the scientists wrote in the *Journal of Experimental Biology*. The chemically addled spider was helpless to resist its zombifier and obediently constructed a special web hammock for the late-instar wasp larva's cocoon, taking about 10 hours to finish one.

Analysis of these cocoon webs revealed that they resembled so-called resting webs that the spiders built when they were ready to molt, shedding their exoskeletons. Such webs lacked sticky, insect-catching spirals. They were also more durable than prey-nabbing webs—a necessary feature for molting spiders, which are helpless and unable to tend to their webs until their molt was over (usually no more than a few hours later). What's more, when healthy and nonparasitized *C. argenteoalba* spiders build resting webs, they embellish the radial strands with special silk structures called fibrous thread decorations, or FTDs. Parasitized spiders also incorporated silk structures that resembled FTDs into their cocoon webs, the researchers found. When they analyzed FTDs in both types of webs, they found that the structures reflected only ultraviolet light. This made the web more visible to fliers capable of seeing light in the ultraviolet spectrum, such as birds or large insects, which might otherwise blunder into a web and destroy it.

In spiders, molting and construction of molting webs were previously found to be triggered when a spider's body was flooded with the release of hormones called ecdysteroids. Perhaps, the Japanese researchers hypothesized, *R. nielseni* wasps were manipulating spiders' internal chemistry by dosing them in a way that mimicked the hormonal release usually associated with molting, "causing the onset of resting web construction," the scientists said.[7]

The result was a spider compelled to build a type of web that would be useful for protecting a wasp cocoon. Threads in the cocoon webs were harder to break than the threads in resting webs because they were produced by repeated weaving actions that were absent in other types of web-spinning. Repetitive weaving in a parasitized spider might be caused by a highly concentrated dose of the wasp's zombifying cocktail.

Other researchers from Brazil soon pinpointed a "smoking gun" (hormonally speaking) for that chemical manipulation. The scientists knew that the ecdysteroid hormone 20-hydroxyecdysone was abundant in molting spiders, as it helped to reinforce their tender new exoskeletons after sloughing off the old ones. So they went looking for signs of the hormone in tiny *Cyclosa* orb-weaver spiders: in nonparasitized spiders that built normal molting webs, and in spiders that were parasitized by wasps in the *Polysphincta* genus and showed signs of unusual web-spinning behavior.

The two spider species—*C. fililineata* and *C. morretes*—both produced resting webs and cocoon webs that resembled each other structurally, and elevated levels of 20-hydroxyecdysone were present in molting spiders and in zombified spiders, the scientists reported in *PLOS ONE*. Steroid levels were highest in spiders that carried third-instar wasp larvae. However, the wasp larvae themselves showed no signs of the ecdysteroid in their bodies. Either the larvae were producing the steroid and injecting it all into the spider—probably during one of their feeding sessions—or they were introducing another type of chemical that induced the spider to make the steroid itself and in far greater quantities than it would while molting.[8]

In 2019, one of the coauthors of that study—Marcelo Oliveira Gonzaga, a professor of ecology at Brazil's Universidade Federal de Uberlândia—collaborated with Eberhard to examine this type of spider zombification more broadly, across multiple species of wasps and spiders.[9] They reviewed dozens of scientific studies about polysphinctine wasps parasitizing a variety of spider species worldwide. Using this data, they first identified different designs for cocoon webs. Next, they fine-tuned modification categories for the cocoon webs, based on details in published photos and in descriptions of the webs. Finally, Eberhard and Gonzaga grouped all the webs into 11 categories based on those modifications, to highlight all the ways that cocoon webs might differ from webs spun by nonparasitized spiders, and then evaluated the hypothesis that wasps manipulate ecdysteroid hormone production in spiders.

There were few published descriptions of what molting webs looked like to use as reference, Eberhard and Gonzaga found, so they supplemented their literature review by conducting new observations of molting webs in five spider species: solitary orb weavers *Allocyclosa bifurca*, *Leucauge mariana*, *Nephila clavipes*, and *Cyrtophora citricola*, and the social spider *Anelosimus eximius*.

They found that designs of cocoon webs often resembled those of molting webs made by nonparasitized spiders. Their observations also confirmed a number of predictions about web construction in the wasps' arachnid victims. Eberhard and Gonzaga discovered that, overall, cocoon webs lacked sticky lines and included reinforced threads that increased the webs' durability. Wasps rarely targeted just one spider species, and when a wasp parasitized multiple species, the spiders' cocoon web designs were not the same; for example, there were 16 different types of cocoon webs associated with wasps in the *Hymenoepimecis* genus, depending on which species of spider they parasitized. When spiders were parasitized by distantly related wasps, the arachnids nonetheless produced similar cocoon webs for their different zombifiers. This suggested that the webs' designs were informed by spiders' genetically encoded mechanisms for building molting webs, which the wasps were likely accessing and hacking, through production or manipulation of ecdysteroids. Whatever psychotropic chemicals the wasps introduced were probably injected into the spider's abdomen, "then carried by the haemolymph to bathe the entire brain," Eberhard and Gonzaga reported.

"Given that wasp larvae presumably produce ecdysteroids to coordinate their own moults, and given that spiders maintain receptors for ecdysteroids throughout the moulting cycle," the study authors wrote, "this breakthrough could have been a relatively small evolutionary step." Wasps may also have evolved additional mechanisms for controlling the amount of ecdysteroids or the timing of their release in the spiders that they parasitize.

Confirming this link will require further research into spider molting hormones, such as introducing ecdysone into a nonparasitized spider to see if that triggers production of a molting web; analyzing and identifying any ecdysone-like compounds that the larvae produce; and then testing those compounds on spiders to see how they affect web production, Eberhard told me in a phone call. Even if ecdysone turns out to be responsible for prompting construction of cocoon webs, it's still unknown if the wasp larvae actively produce ecdysone and dose the spiders, or if instead they chemically induce the spiders to produce and release ecdysone themselves.

“That detail is still not figured out,” Eberhard explains. “There’s still a lot of work to be done in this area.”

*

Wasps have been perfecting their zombifying tricks since the middle of the Triassic period, starting around 240 million years ago. But there’s a group of insects that’s been around even longer: cockroaches. They first appeared nearly 300 million years ago; and while they’d be justified in yelling at young upstart wasps to get off their lawn, they’re still vulnerable to wasps’ manipulation.

Cockroaches are in the insect order Blattodea (along with termites), and there are seven cockroach families containing approximately 4,600 species (we humans typically interact with just a handful of these in the places where we live and work). Our relationship with most cockroaches, or roaches as they’re often called, isn’t exactly friendly—if you Google “cockroach,” the majority of your results will probably be recommendations about how to kill them. In fact, you might be thinking you could never imagine a circumstance in which you would feel sorry for a cockroach. Then again, you’ve probably never seen one under the spell of a tiny wasp called *Ampulex compressa*, also known as the emerald cockroach wasp or jewel wasp.

Those common names aren’t exaggerations; the wasps are brilliantly colorful, and their iridescent greenish-blue bodies are certainly jewel-like. They come from a family of cockroach-parasitizing wasps called Ampulicidae, which includes about 200 species, and they’re found worldwide, mostly in tropical habitats.

Jewel wasps that parasitize the American cockroach (*Periplaneta americana*) hijack roaches’ bodies and turn them into immobile, living meals for their growing larvae. But a jewel wasp's zombification strategy differs somewhat from those of parasitoid wasps that we've met so far. Zombifiers of spiders and caterpillars typically begin by selecting a living host, perhaps briefly paralyzing it, and then laying an egg (or many eggs) on or in its body; dramatic changes to host behavior typically don't kick in until the wasp larvae are ready to pupate.

A jewel wasp mom-to-be, on the other hand, needs to create a compliant cockroach zombie that will willingly follow its wasp mistress to a snug nesting spot, where the roach will placidly submit to being entombed and left to twiddle its antenna, as the wasp young hatch and devour it alive over at least a week. For jewel wasps, there are a few extra steps that they need to take in order to thoroughly zombify a cockroach host, and all of them happen before the wasps commence with the egg-laying part of the program. First on the list are two stings with a modified ovipositor: one to the roach's thorax and one into its head (see figure 9).

At about 1.6 inches (40 millimeters) long on average, American cockroaches are nearly twice the size of their would-be

Figure 9 A jewel wasp (*Ampulex compressa*) attacking an American cockroach (*Periplaneta americana*). *Courtesy of Ken Catania*

zombifiers (female jewel wasps measure about 0.9 inches, or 22 millimeters), but the wasps have speed and agility on their side. A sting under one of the roach's legs on its thorax paralyzes its front limbs for a few minutes. As the roach's legs wobble, the wasp darts in to stab it in the head and deliver potent zombifying compounds, triggering about 30 minutes of vigorous grooming. While the dazed and distracted roach cleans itself up, the wasp is free to make any necessary final preparations for its nest. This cozy spot will also be the cockroach's tomb.

Once the burst of grooming energy wears off, the roach enters a sluggish and docile zombie state. The roach's motor abilities aren't directly affected—it can still walk under its own power and will do so if prodded or led. But while its muscles are still working, its "will" to walk has been disrupted. Other types of locomotion are unaffected; a stung cockroach will swim if dropped in water and can right itself if flipped onto its back—it just won't walk away on its own. This state of diminished movement is known as hypokinesia.

So submissive and lethargic is the now-zombified cockroach that it doesn't even flinch when the wasp bites off the roach's antennae and uses them as straws to slurp its victim's hemolymph. Refreshed by this energizing pick-me-up, the wasp then grasps one of the antennae stumps and walks backward toward its nesting location, leading the roach like a dog on a leash. At the nest, the wasp lays an egg measuring about 0.08 inches (2 millimeters) long, glues it to the cockroach's leg and then seals up the nest with debris, leaving the zombie roach and egg alone together. By now, about an hour has passed since the roach was stung, but it still doesn't try to escape its fate; it passively waits in the darkened nest for whatever horrors are yet to come.

When the egg hatches after another few days, the wasp larva chews a tiny hole in its stoic companion's exoskeleton and begins sipping its hemolymph, treating the zombie roach as a living juice box. Another week or so passes. The larva, now in its second instar, is ready for more solid fare, so it burrows inside the still-living roach to snack on its body tissues and organs.[10] During the third and final instar, the larva first devours the cockroach's fat and muscle tissue; respiratory and reproductive organs are next on the menu, followed by the central nervous system, though the gut and much of the ventral nerve cord are usually left over.[11]

At this point, the roach finally expires; the larva then pupates inside the hollowed-out corpse. About five weeks later, a mature jewel wasp bursts from the cockroach husk and leaves the nest. If the wasp is female, it will soon seek out a fresh roach to zombify.

For more than two decades, Frederic Libersat, a professor at Ben-Gurion University in Israel, has been studying how jewel wasps manipulate cockroaches. In the early 2000s, Libersat and fellow Ben-Gurion University biologists Gal Haspel and Lior Ann Rosenberg made a breakthrough discovery. They found that jewel wasps didn't simply release a payload of zombifying compounds willy-nilly into a cockroach's head. Rather, a wasp injected its venom directly into the roach's brain, deftly bypassing the blood-brain barrier and targeting precise locations in the central nervous system to mute the insect's normal escape reflex.

To track the pathways of this venom delivery, the scientists modified the venom in several wasps. They added radioactive amino acids, then funneled the "doctored" venom back into jewel wasps. After allowing those wasps to sting cockroaches, the researchers then looked for traces of radioactivity in the roaches' heads. The scientists found that radioactive levels were higher in the cockroaches' cerebral ganglia, or brain tissue, than in the surrounding tissues. Specifically: the subesophageal ganglion (SEG) and the larger supraesophageal ganglion (SupEG).

A jewel wasp's stinger measures about 0.1 inch (2.5 millimeters) long. While that may not sound like much, it's plenty long enough to poke into the ganglia, which lie inside the roach's head at a depth of 0.04 to 0.08 inches (1 to 2 millimeters).

"To our knowledge, this is the first direct evidence documenting targeted delivery of venom by a predator into the brain of its prey," Haspel, Rosenberg, and Libersat reported in the *Journal of Neurobiology*.[12]

Jewel wasps likely use special sensors on their stingers to locate the ganglia for their zombifying injection, the three researchers later learned through another series of experiments. When they presented wasps with cockroaches in which the cerebral ganglia had been surgically removed, the wasps fumbled to locate the missing targets. Stinging the roaches' heads, normally a swift, in-and-out operation, took 15 times longer when the roaches had no brains.[13]

Other studies had previously suggested that the SEG in cockroaches was associated with triggering locomotion. If that were true, Libersat and biologist Ram Gal hypothesized in *PLOS ONE*, perhaps compounds in the jewel wasp's venom were muting activity in the SEG and preventing the roach from initiating movement, so it would only walk when a wasp was leading it.

In their study, Gal and Libersat looked for evidence connecting SEG activity—or lack of it—with spontaneous walking.[14] They measured brain activity in roaches using microelectrodes, noting that stung roaches showed less electrical activity in their SEG. They took cockroaches that hadn't been stung and injected them in the SEG with venom milked from wasps. Like zombified wasps, those roaches were less inclined to walk on their own. Inhibiting neuronal activity in cockroaches' SEG with an anesthetic also reduced their motivation for unprompted walking.

"Our data unequivocally demonstrate the role of the SEG in the venom-induced inhibition of the drive for walking in cockroaches stung by *A. compressa*," providing the first direct evidence of the connection, Gal and Libersat wrote. "Given these facts, one can only wonder how, through millions of years of coevolution, *Ampulex* has evolved this exquisite strategy to chemically control such 'higher' behavioral function in its host."

With precision tools and potent chemical weapons, jewel wasps seem unbeatable. However, cockroaches aren't entirely defenseless and have been known to fight back—and sometimes best their attackers. Scuffles between wasp and roach were vividly described in 1942 by American entomologist Francis Xavier Williams, who captured and observed the wasps while surveying insects in Nouméa, New Caledonia, in the southwest Pacific Ocean.

A roach under attack "struggles furiously, twisting, straining and describing short jerky circles, parrying with its legs, and striving particularly to tuck in its chin so that the tenaciously clinging wasp will not sting its throat," Williams reported in *Proceedings of the Hawaiian Entomological Society*. While healthy, young, female wasps were usually more than a match for their prey, "an old or incautious wasp attacking a vigorous and aggressive roach may be badly bitten in the abdomen, the roach's jaws even breaking a tough sclerite so that the wasp succumbs in a few days," Williams said.[15]

Biologist Kenneth C. Catania, a professor at Vanderbilt University in Tennessee, studies animal sensory systems, and he suspected that the relationship between jewel wasps and roaches could be a unique opportunity to not only document the wasp's attack strategy but also analyze the cockroach's defensive moves. To do that, Catania used something that was unavailable to Williams in 1942: high-speed video, recorded at 1,000 frames per second. Highly magnified, slow-motion footage can reveal details of insect interactions that are impossible to see with the naked eye.

Catania published his findings in 2018 in the journal *Brain, Behavior and Evolution*, describing data collected from 55 video trials that pitted cockroaches against would-be zombifiers. Its title: "How Not to Be Turned into a Zombie."[16]

In Catania's trials, 28 cockroaches—about half of the group—seemed blissfully unaware of the wasps until they were attacked.

Only four of those roaches were able to evade being stung after a wasp grabbed them.

But 27 roaches began taking defensive actions early, while the wasps were still stalking them. These wary roaches would "stilt-stand"—straightening their legs to elevate their bodies. They would angle themselves so that their rear legs were oriented toward the wasps. If a wasp touched a roach with an exploratory leg or antenna, the roach would kick it. These powerful kicks "often sent the wasp careening into the walls of the filming chamber," Catania reported. He described the kicks as similar to the swing of a baseball bat. There was a windup, a swing, and a follow-through arc that incorporated the roach's entire body.

Roaches also defended themselves against the wasps' lunge attacks. The roaches turned their bodies, scraped at wasps with their leg spines, held the wasps off with a so-called stiff-arm defense, and bit at their attackers' heads and limbs.

Not all these defenses were successful, but 63 percent of the cockroaches did manage to escape a sting (or were at least able to hold off the attacker for three minutes, whereupon the timer ran out and the experiment ended). And some wasps appeared to give up when confronted with a cockroach that fought back.

From this investigation comes a new perspective on cockroaches' chances against wasp zombifiers and a glimpse of the defensive strategy that works best for them, Catania wrote: "Be vigilant, protect your throat, and strike repeatedly at the head of the attacker."

*

Being able to fight your way out of potential zombification is certainly a big plus in the survival game. But as anyone in the middle of a zombie apocalypse knows, subduing your attacker won't get you too far if it manages to infect you. In stories about weathering an undead Armageddon, no moment is more poignant than when a character defeats a zombie and thinks they're home free, only to realize that they've been bitten, scratched, or otherwise compromised and are now doomed to a fate worse than death. In a nail-biter of a scene from HBO's *The Last of Us*, a pregnant Anna (Ashley Johnson) fights for her life against an "Infected" attacker while in labor; she kills her assailant and gives birth—only to notice a bite on her leg. Anna then begs her friend Marlene (Merle Dandridge) to kill her and take her newborn away. They both know that it's too late for Anna—she can't be helped or saved.

Anna's baby grows up to be Ellie (Bella Ramsey), who harbors a unique immunity to the fungus's mind-controlling manipulation. But for everyone else in that world—and in most zombie apocalypse scenarios—infection is a one-way ticket to zombietown. No matter which method transmits a zombifying pathogen, once that happens, it's just a matter of time before the victim is completely and irreversibly overtaken.

Such is also the case for zombifying wasps' invertebrate victims—most of the time. But there's a notable exception: beetle zombies created by the braconid wasp species, *Dinocampus coccinellae.*

D. coccinellae, or green-eyed wasps, are exceptionally widespread, found worldwide in temperate, subtropical and tropical locations. Victims of green-eyed wasps include at least 72 species of ladybugs (also known as ladybirds or ladybeetles, in the beetle family Coccinellidae), which may be parasitized as pupae, larvae, or adults.[17] Implanted eggs hatch wasp larvae that burst from adult ladybugs' bodies, then coerce the ladybugs into actively guarding their cocoons when the larvae enter metamorphosis. But this zombification isn't necessarily a death sentence, or even a permanent state. Unlike the cockroaches, spiders, and caterpillars that fall prey to manipulation by other parasitoid wasps, ladybug victims of *D. coccinellae* are known to survive and recover from the ordeal.

A green-eyed wasp larva that hatches inside an adult ladybug feeds on the beetle's hemolymph and grows for about 20 days, then squeezes, squirms, and wriggles its yellow-white cylindrical body (which by this point is nearly as big as the ladybug's) out of its host's abdomen. This is when behavior manipulation of the ladybug begins, researchers in Canada and France reported in *Biology Letters*. The larva spins a cocoon between its host's legs, and the beetle—which appears partially paralyzed—holds fast to the silk pod. It doesn't stray from its post, nor does it eat (though it "twitches at irregular intervals") until the adult wasp emerges about seven days later, the study authors wrote. They found that the presence of a living, twitching ladybug guardian on a wasp's cocoon helped keep it safe from predators; only about 33 percent of protected cocoons fell victim to predators, compared with 85 percent of cocoons that were clutched by dead ladybugs. And surprisingly, ladybug zombification was reversible; in experiments, about 25 percent of the manipulated ladybugs recovered once they were released from their bodyguard responsibilities.[18]

A few years after that study was published, several of its coauthors uncovered another piece of the ladybug zombification puzzle. They analyzed gene activity in parasitized ladybugs, searching for unfamiliar genetic material. Foreign viral RNA in the ladybugs' brains hinted that *D. coccinellae* was using a remote-controlled secret weapon: a virus.

D. coccinellae paralysis virus, or DcPV, as the study authors named it, is stored by female wasps in their oviducts and is delivered into hosts along with implanted eggs, the researchers reported in *Proceedings of the Royal Society B*. As a larva grows inside its host, DcPV replicates in the ladybug's brain and nervous tissues. At the same time, certain immunity-related genes in the ladybug are suppressed, so that the wasp intruder can grow undisturbed. When the larva finally wriggles out, the researchers proposed, its departure may abruptly trigger an immune response to the accumulation of viral particles in the ladybug's cerebral ganglia, which could induce the paralysis displayed by the zombified guardians. When the virus is cleared from a ladybug's system, the insect would often resume normal behavior, including feeding and reproduction, the study authors reported.[19]

When this viral partnership was first discovered, it was a source of great excitement for lead author Nolwenn M. Dheilly, now an assistant professor at Stony Brook University in New York (Dheilly led the new study while at the University of Perpignan in France). Dheilly reacted at the time by running around her lab, exclaiming, "I've got a virus! There's a virus in the head of the lady beetle!"[20]

Outside of scientific laboratories and field sites, examples of insect zombification rarely spark this level of enthusiasm. But one zombified ladybug's plight recently captivated millions, after videos of the mind-controlled beetle went viral on TikTok. Content creator Tiana Gayton is so enamored of insects and spiders that she features them regularly in her numerous TikTok videos, some of which garner millions of views. Her unexpected discovery of a zombie happened in April 2021 at a grocery store, where Gayton found a ladybug crouching on a bunch of organic cilantro.

"I noticed she was guarding something," Gayton said in the video (identifying the ladybug as a female was a subjective choice; at first glance, male and female ladybugs look much the same, though females are usually bigger). Closer inspection revealed that the ladybug was grasping a cocoon; a quick consultation with Google told Gayton that the ladybug was probably under the control of a parasitoid wasp. Naturally, Gayton decided to rescue it. She documented the process step-by-step for her rapt viewers, many of whom left comments about the bug's cuteness, and they enthusiastically rooted for the insect's recovery.

Using a toothpick, Gayton gently pried the ladybug's legs away from the tangle of silk around the cocoon, then transferred the freed insect into a jar. Gayton named the beetle "LadyBerry," feeding it sugar water, blueberries, banana, water-soaked raisins, and eventually aphids (which she gave to the ladybug on camera, much to her viewers' collective delight). LadyBerry was lethargic at first but slowly became more active, crawling around her new home, flexing her wings, and avidly devouring her meals.

"She began to wake up," Gayton said. "It was a whole new ladybug."

About a month after her rescue, LadyBerry was strong enough to fly. Gayton knew that it was time to say goodbye, and she released the tiny TikTok star back into the wild (a local park).

"It was nice knowing you, LadyBerry," Gayton said in a heartfelt farewell video, as the former zombie spread her wings and flew away.[21]

A ladybug's average lifespan is one to two years, so it's possible that LadyBerry went on to enjoy many more months of post-parasitized, buggy bliss. No matter what happened to LadyBerry, for Gayton—and for the millions of viewers who followed her saga—she'll be remembered as the zombie that beat the odds and achieved something that most zombies never get: a second chance.

*

Even as researchers delve further into the habits and behaviors of known ichneumonid wasps, hundreds of new species are being discovered and identified. The number of described Braconidae wasp species in Australia alone more than doubled in 2018—from 99 species to 236 species—thanks to DNA barcoding, a technique that identifies organisms using short segments from specific regions of their genetic code. One newly identified Australian wasp, described in the *Journal of Hymenoptera*

Research, had a "remarkably long ovipositor" compared to those of other species in its genus, the researchers wrote. That unusual structure—along with its reproductive similarities to the monstrous "xenomorph" from the movie *Alien*—prompted scientists to name the wasp *Dolichogenidea xenomorph*.[22]

"At less than 5 mm in length, *Dolichogenidea xenomorph* might seem to lack the punch of its fearsome namesake," said lead study author Erinn Fagan-Jeffries, a postdoctoral researcher in the University of Adelaide's School of Biological Sciences, "but size is relative." Fagan-Jeffries added, "To a host caterpillar, it's an awesome predator."[23]

There are plenty of life-cycle and behavioral similarities between the fictional *Alien* xenomorph and parasitoid wasps, but did the creators of *Alien* design the xenomorph's habits based on a wasp's? For this film nerd, *Alien* is a beloved classic that I've watched more times than I can count and, yes, the chest-bursting scene still gives me chills. Perhaps it was time for me to get to the bottom of how its creators got the idea for their extraterrestrial's parasitic life stage. Before that electrifying scene existed on film, it was words on a page. I needed to backtrack to explore the motivations of *Alien*'s screenwriter: the late Dan O'Bannon, who died in 2009.

An obituary in the *Los Angeles Times* hinted that the young xenomorph's eruption from Kane's body was written as a nod to parasitic lifestyles in general, if not parasitoid wasps specifically. The obituary recalled a 2003 interview in Newark's *Star-Ledger*, in which O'Bannon mentioned that he saw the scene as a chance to infuse his xenomorph with some of the more gruesome traits found in actual parasites (though, he didn't specify which parasites those were).

"The real world offered many examples which were extremely loathsome," O'Bannon said. "I thought, if it's good enough for Mother Nature, maybe it will work on an audience."[24]

However, other accounts have hinted that the real inspiration for O'Bannon's chest-burster was a little more personal. The idea of a creature violently thrashing around and tearing its host apart from the inside may have sprung from O'Bannon's long-standing struggle with intense and chronic stomach pain. His condition was eventually identified in 1980 as Crohn's disease, according to American film critic Jason Zinoman. Crohn's is an inflammatory bowel disease that leads to swelling of the digestive tract; the large intestine or small intestine—or both—may be affected. Complications can be life-threatening, and while there are mitigating medications and therapies, there is no known cure.

Zinoman wrote that O'Bannon suffered from debilitating digestive distress for years before it was diagnosed; the *New York Times* reported in 2009 that it eventually caused O'Bannon's death.[25] Crohn's is a stressful illness, as it makes eating (and the aftermath) incredibly painful. That pain and stress, Zinoman hinted, may have planted the seeds for the shocking scene.

"The digestion process felt like something bubbling inside of him struggling to get out," Zinoman wrote. "What he didn't realize back then was that this lifelong struggle would actually be the inspiration for his greatest idea."[26]

It was looking like parasitoid wasps didn't shape the creation of *Alien*'s deadly extraterrestrial predator after all. But then I learned that James Cameron, director of *Aliens*—the second film in the franchise—admitted to a personal experience that connected wasps to the xenomorph in a different way.

During a chilling climactic scene in *Aliens*, Ellen Ripley (Sigourney Weaver) is on a mission to rescue a young girl from an exomoon colony decimated by xenomorphs. She finds the child, only to then stumble into a chamber that's densely packed with xenomorph eggs. As Ripley stands frozen in the darkened room, a loud squelching sound alerts her that an enormous insect-like xenomorph queen is busy laying those eggs nearby, flanked by smaller xenomorph offspring. Cameron said that during those first electrifying moments, when Ripley is near-paralyzed with terror, he wanted to capture what he'd felt during a vivid nightmare in which he was inside a room in which the walls and ceilings were completely covered with crawling masses of wasps.

"If I moved, if I breathed, if I did anything, they would attack," Cameron said. But the swarm of dream wasps wasn't attacking him—at least, not yet. "There was this moment to appreciate exactly how screwed I was," he recalled. "So, I remember that feeling and I thought, 'How can I translate that feeling to the moment in a horror film?' "[27] Even if those wasps in Cameron's nightmare weren't the tiny parasitoids that target and manipulate caterpillars, spiders, roaches, and ladybugs, something about wasps got into Cameron's head—with unforgettable results for xenomorph movie fans everywhere.

Parasitic wasps do play deadly mind games with their victims, though most of the time they don't literally set up shop inside their hosts' heads to manipulate behavior. As it happens, there are other parasitic insects that regularly get up close and personal with their zombified host's brain. For some of these headcases, zombifying an invertebrate victim culminates in a spectacular head-splitting finale.

8

ANT DECAPITATORS AND THE RISE OF THE ZOMBEES

Zombifying Flies

"A newly infected zombie is almost impossible to detect."

—COLONEL DARCY (SAM RILEY), *Pride and Prejudice and Zombies*

The persistent buzzing of flies in a sickroom is never a good sign. Flies are typically associated with decay and death, and their presence near a patient hints that the person is probably a lot worse off than they seem. This is especially true in the film *Pride and Prejudice and Zombies*. The movie, which reworks Jane Austen's classic story of Regency-era British upper-class romance to include a plague of the undead, features flies as an important zombie-detecting tool used by the ruthless Colonel Darcy (Sam Riley). Darcy carries a small glass vial of living flies, which he deploys like tiny canaries in a zombie-infested coal mine, to detect if a person who might appear normal is, in reality, a zombie. When Darcy releases a few of the insects near the bed of a desperately ill Jane Bennet (Bella Heathcote), whom he suspects has been bitten and infected, he expects the flies to be attracted to her "dead" flesh and reveal her zombie status. But Jane's sister Elizabeth (Lily James), determined to protect her sibling—and an adept zombie-fighter in her own right—snatches the flies out of the air one by one and grimly returns them to Darcy as a handful of pulp. (His suspicions turned out to be misplaced: a bite-like injury on Jane's hand was caused by a backfiring musket, and she was ill with the flu, not a zombie infection.)

For those of us who aren't zombie-hunters (and for plenty of animals, too) flies are generally seen as an annoying nuisance and a potential health hazard, to be swatted or smushed at all times. They buzz in your ears, dive-bomb your trash, prance all over your food, and gather on eyes, open wounds, or sores if given the chance. They feed and breed in gore, decaying matter, feces, and trash, and their habits of pooping, drooling, and barfing wherever they land make them vectors for dozens of diseases.

The fly that people are most intimately acquainted with is generally the house fly (*Musca domestica*), the most common and widespread fly species. House flies have been around for about 65 million years, originating in the Eastern Hemisphere and eventually dispersing to live everywhere in the world, especially in places where there are people and livestock. But while *M. domestica* is probably the species that you picture when you hear the word "fly," it's just one species in an insect group that is vast and incredibly diverse. And while Colonel Darcy in this alternate *Pride and Prejudice* universe used flies to detect zombies, some flies are more interested in creating them.

All flies belong to the order Diptera. The name comes from the Greek words *di* ("two") and *ptera* ("wings"), and insects in this order, commonly known as the "true flies," have just two wings (most other insects have four; remnants of dipterans' rear wings exist as modified barbell-like structures called halteres, which help stabilize the insects during takeoff and flight). Diptera includes approximately 150,000 described species in 10,000 genera—which means that about 14 percent of the world's insects are flies—and there are probably many more species of true flies that are yet to be identified, according to the University of British Columbia.[1] Accompanying their signature single pair of wings, flies typically have large, compound eyes, soft bodies, and mouthparts specialized for piercing, sucking, lapping and absorbing liquids, rather than biting and chewing. All produce legless larvae, which are often called maggots. Unlike adult flies, larvae have mouth hooks or chewing mouthparts.

Alongside house flies, Diptera includes fruit flies, midges, gnats, crane flies, bot flies, blow flies, bottle flies, and mosquitoes, to name just a few. The biggest Diptera species is the fly *Gauromydas heros*, which is known from a handful of sightings in Brazil and can grow to measure a whopping 2.8 inches (7 centimeters) long. On the other end of the fly size spectrum is the tiny *Euryplatea nanaknihali*, which is easily dwarfed by a single grain of sand. With a body measuring just 0.0157 inches (0.4 millimeters) long, *E. nanaknihali* is one-third the size of the only other species in its genus, *Euryplatea eidmanni*. It lives in Thailand and was described in 2012 by Brian Brown, curator emeritus of entomology at the Natural History Museum of Los Angeles County, who identified the insect as "the smallest known fly in the world."[2] Brown later described another contender for the title of "world's smallest fly" in 2018, with the discovery of *Megapropodiphora arnoldi*. A single *M. arnoldi* specimen, found at a site near Manaus in northern Brazil, measured a mere 0.0156 inches (0.395 millimeters)—just a hair smaller than *E. nanaknihali*.

"With the discovery of this species, flies move into the 'nanosphere' of insect life," Brown wrote.[3]

Why would these flies need to be so very tiny? So they can develop as larvae inside the wee heads of only slightly less tiny ants.

Minuscule *M. arnoldi* and *E. nanaknihali* both belong to the fly family Phoridae and are known as phorid flies. Phorids are sometimes referred to as scuttle flies for their erratic scurrying movements; they are also known as coffin flies for their habits of breeding in cemeteries and mausoleums, and are called humpbacked flies for their hunched bodies. They thrive everywhere on Earth but in the coldest and driest habitats, and there are more than 4,000 phorid species distributed around the world (though, by some estimates this may represent less than 10 percent of the group's true diversity).[4]

"Phorids are one of the most underappreciated groups of insects," says Brown, who has studied phorids for decades. Most people, if they think of phorids at all, lump them all together as scavengers "because those species are really common and ubiquitous worldwide," he explains. But phorids have a broad array of lifestyles—including parasitism. There are also phorid herbivores and fungivores, in addition to the better-known scavengers.

Phorid flies are all exceptionally small—the biggest phorids are no more than 0.2 inches (6 millimeters) long—and some have extreme modifications to their bodies that make them resemble termites or ant larvae, so they can infiltrate the colonies of insects that they parasitize.

The majority of phorid species that people encounter on a day-to-day basis reproduce by laying their eggs in decaying vegetation or other organic matter, but there are many phorid flies that prefer their eggs to hatch and grow in sustenance a bit fresher than that. For these flies, only the body of a living insect host will do, and these very tiny parasitoid flies can trigger big changes in the behaviors of their victims. The mere appearance of phorid flies can dramatically disrupt foraging activity in an entire ant colony, opening a window of opportunity for rival colonies to move in and take over. Ants that fall victim to phorids soon find that their minds are no longer their own. Eventually, neither are their heads.

Phorid species across several genera, including those of the tiniest "nano" species that Brown described, target ants, and they have a special strategy for giving their larvae a head start in life. As for how this method turns out for the ant, the nickname for this group of phorids—"ant-decapitating flies"—should tell you all you need to know.

Scientists first described the gruesome habits of ant decapitators in the phorid genus *Pseudacteon* more than 90 years ago, from observations among ant populations in Europe, South America, and the United States.[5] A female fly begins by staking out a worker ant—carefully keeping her distance at first, as she is no bigger than her target's head. To scientists observing phorids in the field, "they appear as minute, fuzzy specks as they hover over host ants," Lloyd Morrison, an ecologist for the National Park Service, wrote in a guide to insect parasitoids in North America.[6] Females don't have a lot of time to be choosy about their hosts, as adult flies live only for about a week or less in the wild.

When the female fly sees an opening, she darts in and lays an egg inside the ant's thorax—one and done—in less than a second (analysis of the female reproductive system in the phorid fly *Pseudacteon wasmanni* revealed that eggs are torpedo-shaped and measure 130 micrometers long, or about 0.005 inches).[7] A single *Pseudacteon* female can produce from 200 to nearly 300 eggs, and in a single hour she may make more than 100 parasitizing attempts (though she only lays one egg per host). Newly parasitized workers "frequently appear stunned after an oviposition strike," US Department of Agriculture entomologist Sanford Porter wrote in *Florida Entomologist*, and the ants "often stilt up on their legs for a few seconds to a minute before running away."[8]

These egg-laying attempts don't all succeed; indeed most of them flop. In laboratory experiments, when *Pseudacteon* females tried to implant an egg in an unwilling ant, they failed at least 65 percent of the time.[9] But when an egg does manage to end up inside an ant, its host enters the realm of the walking dead. Once the egg hatches, the ant has only a few weeks of life before it succumbs to the manipulations of its attacker, stumbling away from its home and family and then undergoing decapitation from the inside out.

Within days after hatching, the phorid larva migrates from the thorax into the ant's head; little is known about how the parasitoid avoids being destroyed by the ant's immune system, but one possibility is that moving quickly into the host's head may help the larva evade an immune response.[10] For the duration of the larva's second instar—about two to three weeks—it makes itself comfortable in the ant's head cavity, sipping on hemolymph. This liquid nutrition is all the larva needs until it reaches its third instar.

For an infected ant during those initial honeymoon weeks, despite carrying and nourishing a growing parasite inside its head, life is pretty much business as usual; the ant looks and behaves normally, according to scientists with the Louisiana State University (LSU) College of Agriculture. Through "intensive observation" of ant parasitism by the phorid fly *Pseudacteon tricuspis*, LSU researchers found that a parasitized ant stayed with its nestmates until about 8 to 10 hours before the larva in its head was ready to pupate. It would then depart the nest on what appeared to be a normal foraging expedition, alongside its non-parasitized sisters. But for the zombified ant, this final excursion was a one-way trip. Once the ant turned its back on the colony and walked away, it was on a death march. And the parasite was in the driver's seat.

"Parasitized ants were highly mobile after they left the nest and ultimately entered the thatch layer at the soil surface," the scientists reported. "The term 'zombie' fire ant workers was coined to characterize the behavior while under parasitoid control."[11]

Finally, the phorid larva is ready for its metamorphosis. It releases an enzyme that degrades membranes in the wandering ant's exoskeleton, causing the ant to stop walking and eventually collapse. The ant's head loosens from the body, as does the first pair of legs; other legs may be affected, too. Its mandibles weaken, rendering it unable to bite or burrow.

As for the larva, it indulges a new appetite for solid food; namely, ant head tissue.

You can probably guess where this is headed; the ant's hollowed-out, larvae-stuffed head falls off (the ant, unsurprisingly, is already dead by now, even though its legs are often still twitching as its head rolls away). The parasitoid, however, is just fine. It finishes off the last of the tasty bits inside the decapitated ant head and pushes the mandibles out of the way—the ant's no longer using them, after all—and then the larva wriggles into position so that its first three pupal segments are stuffed into the gap where the ant's mouthparts used to be. These segments harden and darken, becoming a tough, protective plate that's roughly the same color as the ant's exoskeleton, and two hairlike breathing "horns" extend from the pupa out on either side of the ant's mouth opening.

Other parasitoid insects keep their zombie host alive until the larva's metamorphosis is over, but phorids pupate unguarded inside their dead hosts' disembodied heads. However, the exoskeleton of an ant's head is extremely hard—tougher than other parts of its body—and therefore lends extra protection to the pupating larva, Brown says.

Two to six weeks later, depending on air temperature and species size, the adult phorid fly is ready to pop out from inside the detached ant head (see figure 10), like the goddess Athena of legend springing fully grown from the head of her father, Zeus. Only, this newborn is a lot smaller than an ancient Greek deity and has more legs than most. A few hours after emerging, the adult phorid fly is ready to mate—and continue its head-splitting reproductive cycle.

Parasitized ants seemed to be acting under the sway of the larvae "in a way that benefits the survival of the parasitoid and ultimately the adult fly"—an important hallmark of zombification—

Figure 10 A newly hatched phorid fly emerges from the decapitated head of a red imported fire ant (*Solenopsis invicta*) that the fly parasitized and killed. *Courtesy of the USDA Agricultural Research Service*

LSU Agricultural Center entomologists Don C. Henne and Seth J. Johnson reported in *Insectes Sociaux*. Henne and Johnson observed the phorid species *P. tricuspis* as the flies parasitized *Solenopsis invicta* ants (also known as red fire ants, more about this phorid–ant relationship shortly). In their laboratory observations, Henne and Johnson noted that parasitized ants spent most of their time tending to the colony's brood and didn't go out to forage much. By staying in the nest until it was time for the phorid to pupate, an ant host kept itself—and its larva passenger—safe from predators and from other parasites, so it was likely that the larva was manipulating the ant into behaving this way.[12]

Leaving the nest prior to pupation is also helpful for the parasitoid, as a dead ant in the nest would be quickly cleared away by workers and tossed in the colony's trash heap, where the corpse and its pupa passenger could be eaten by scavengers. Prior research referenced by Henne and Johnson suggested that the brain was the very last bit of head tissue that the larva devoured, hinting that as a parasitized ant staggered from its home for the last time, it still had a (mostly) functioning brain. This would also benefit the larva, which at that point would still require the zombified ant to at least have a working sensory system, in order for it to walk away from the nest under its own power and find a safe spot for the larva to pupate.

Up to that point, it's likely important for the ant to at least go through the motions of normal behavior. Social insects like ants are quick to respond if one of their kin appears to be infected, as the spread of parasites or disease pose a significant threat to the colony, and ants may kill nestmates or evict them if they behave erratically. Little is known about how phorid zombifiers keep their presence hidden from their host's nestmates until they prod their host into leaving the nest, but they may not always succeed in their subterfuge, says Brown. He recalls an instance in Costa Rica when he saw a group of leaf cutter ants that had pinned down one of their major workers—the caste of workers that usually serve as the nest's defenders—and were pulling it to pieces.

"I thought, wow, that's weird," he says. His curiosity piqued, Brown picked up the dead worker and pried open its head. Sure enough, inside was a phorid larva in the *Neodohrniphora* genus.

"To me that indicates that some ants can tell that their sisters are parasitized," he says. "In that case, they [the parasite] weren't successful in hiding—but I don't know how common that is."

*

Parasitized fire ants aren't the only victims that fall under the phorids' spell. The mere presence of phorid flies buzzing around a fire ant nest is enough to dramatically change the colony's behavior, affecting ants that aren't directly parasitized.

There are over 70 species of ant-attacking *Pseudacteon* phorid flies, and at least 24 of those in South America are known to attack ants across multiple species in the *Solenopsis* genus. While there are more than 190 *Solenopsis* ant species worldwide, the most infamous is a group of about 20 species, mostly from South America, known as fire ants. These ants get their name from the burning sensation that accompanies their venomous stings; as they bite down and cling tightly to an attacker with their mandibles, they pivot their bodies to repeatedly stab and inject venom with stingers on their rear ends. If a fire ant nest is disturbed, hundreds of insects quickly mobilize into a stinging swarm, and together the insects can deliver enough venom to be fatal to small or young animals. In humans, fire ant stings typically cause painful and itchy welts, blisters, and fluid-filled bumps that can last a week or more. Their venom is irritating but usually not life-threatening to people, though about 2 percent of the time fire ant stings can cause serious allergic reactions.[13]

Over the past century, fire ants have proven to be highly effective invaders, especially the red fire ant: *Solenopsis invicta.* These reddish-brown ants, which measure no more than 0.25 inches (0.64 centimeters) long, are infamously invasive in the United States, where they arrived by hitchhiking in imported soil or ballast, surfacing in Alabama in the 1930s. *S. invicta* is now firmly established across most of the southern United States and in Puerto Rico and has spread west into California, and has been discovered sporadically as far north as Maryland. Fire ant populations have also sprung up in Australia, elsewhere in the Caribbean, and in China and Mexico. In September 2023, scientists reported in the journal *Current Biology* that they recently counted 88 *S. invicta* nests in Sicily near the city of Syracuse, with each nest containing many thousands of ants. This marks the first time that the invasive species has been identified in Europe, and now that the insects have a foothold there, experts fear that fire ants could spread across the continent.[14]

Fire ant colonies with a single queen, or monogyne colonies, can be vast underground complexes, incorporating up to 150 surface nest mounds per acre and housing roughly 7 million ants.[15] As many as 500,000 workers may be busily foraging at any given time, wandering up to distances of about 300 feet (91 meters) from their nests. Red fire ant colonies can also have multiple fertile queens; such arrangements are known as polygyne colonies. They may contain more than 200 mounds per acre and host up to 40 million ants. Experts with the USDA Agricultural Research Service estimate that fire ants cover about 350 million acres of land across the United States; if you were to sweep all the US's invasive *S. invicta* ants together and shape them into a giant sphere, it would weigh about 13,475 million tons (12,224 million metric tons).[16] In the United States alone, crop loss and other damage caused by red fire ants and their invasive cousins, black fire ants (*Solenopsis richteri*), carry a price tag of approximately $8 billion per year.

Wherever these determined ants invade, they have few natural enemies, so it's relatively easy for them to multiply rapidly. Once established, they outcompete native ant species (even native fire ants, such as *Solenopsis amblychila*, *Solenopsis aurea*, and *Solenopsis xyloni* in North America), disrupt ecosystems, and threaten local wildlife. Fire ant workers also devour young crop plants. And while most of their colony infrastructure is in tunnels underground, jutting surface mounds of loosely packed soil, which can measure several feet tall, can damage farm equipment. The ants have a highly aggressive response to nest disturbances and just walking near a mound can trigger an alarm signal, which creates risks for people of being swarmed and stung.

Fire ants have been known to settle inside houses and other buildings if food is plentiful there, or if the colony had been recently displaced by landscaping, drought, or floods. But while fire ants aren't especially intimidated by humans, the ants become extremely agitated in the presence of tiny, ant-decapitating *Pseudacteon* phorid flies. Parasitoid phorids are thought to be drawn to olfactory cues released by fire ants; the flies follow a chemical cloud of volatiles in a combination of ant venom alkaloids and distress pheromones. In Brazil, when a fire ant nest is disturbed, phorids come calling within 20 minutes. And when phorids show up near a fire ant nest, all hell breaks loose.

"The ants will run and hide, freeze, stop foraging, or twist upside down so the flies don't sting them," Sanford Porter told *AgResearch Magazine*. When phorids drop by, fire ants also drastically change how they collect and store food. They may try to hide food caches, burying them under dirt or debris. They may switch to foraging routes that travel through subterranean tunnels rather than on the surface, or they may forgo daytime expeditions and spend more time foraging at night under the cover of darkness.[17] Scientists also found that when phorids were present, fire ant colonies dispatched smaller worker ants—which were usually assigned to nursery duty—to do the foraging, throwing the colony's critical brood care regimen into disarray.[18]

What's more, *S. invicta* colonies continue to modify their foraging behavior even after visiting phorids leave, Porter and entomologist Li Chen, then an associate professor with the Chinese Academy of Sciences in Beijing, reported in 2020 in *Insects*. As extreme as the ants' reactions are, the flies' actual success rate of parasitizing ants is very low—analysis of fire ant colonies found that only about 1 percent of the ants were actually infected with parasitoid larvae.[19] Nevertheless, the appearance of just a handful of phorid flies is enough to sow chaos among fire ant foragers, and the ant workers' panicked and prolonged response to all phorid visits hints that the flies pose a serious threat to their colonies, Porter says.

"The only way they could have evolved these defensive behaviors is if the flies had some effect on fire ant populations," he told *AgResearch Magazine*.

By disrupting business-as-usual activity in fire ant colonies, phorid flies distract the ants from collecting food, defending their homes, and tending to their young. This offers other ant species a chance to step in where they were formerly displaced or to move into new areas where fire ants previously kept other ants away. Fire ants are far less abundant in their native South American locales than they are in their adopted homes, coexisting alongside many other competitive ant species, and the presence of multiple species of phorid flies there is thought to be at least partly responsible for keeping fire ant numbers in check (there are 10 different fly species that target *S. invicta* in Argentina alone).

Phorids in South America are not only excellent at regulating fire ants; their relationship with their favored host species—the result of millions of years of evolution—is so specific that most parasitoid phorid species typically target just one species of fire ant. For example, there are native phorid species in the United States that prey on North American fire ants, while ignoring the invasive species. Decades after South American fire ants made themselves comfortable across the southern US, a growing body of scientific evidence persuaded the USDA that invasive *S. invicta* might have an Achilles heel: South American phorid species that parasitized *S. invicta* on their home turf.

In the early 1990s, the USDA approved importing *Pseudacteon* phorid species that typically harass and decapitate *S. invicta*. Efforts were led by two research programs: one at the University of Texas in Austin and one under the USDA Agricultural Research Service in Gainesville, Florida. *Pseudacteon tricuspis* was the first phorid the program deployed, with the release of captive-reared flies in Florida in 1997. Since then, 11 states in the southern US have seeded multiple species of *Pseudacteon* flies across numerous sites. Among the introduced phorid species are *P. tricuspis*, *P. litoralis*, *P. obtusus*, and *P. nocens*, which mainly attack larger fire ant workers, and *P. obtusitus*, *P. curvatus*, and *P. cultellatus*, which favor smaller workers.[20]

However, zombifying phorids on their own likely won't be a silver bullet for eradicating invasive ants. Once phorids are introduced, it can take years to determine which flies are the best fit for year-round climate conditions in different regions where fire ants have made themselves at home, Li Chen and Henry Y. Fadamiro, a professor of entomology at Texas A&M University, wrote in the *Annual Review of Entomology*. Even after phorids are matched to a local biome, "populations can take up to four years to reach maximum levels, and native ants require one or two more years for their populations to build up in competition with the imported fire ants," Chen and Fadamiro reported.[21]

Still, some results have been promising. As predicted, imported phorids focus on their preferred ant species—the red fire ant *S. invicta* and its close relative, the black imported fire ant *S. richteri*—and ignore native ants. The phorid species *P. tricuspis* and *P. curvatus* were the first to become established in most locations in the southern US where invasive fire ants are found, and by 2021, "establishment and geographical range expansion have been confirmed for six introduced *Pseudacteon* species in the US," Chen and Lloyd W. Morrison wrote in *Biological Control*. Ultimately, controlling fire ants will likely require a multipronged approach, combining phorid zombifiers and naturally occurring pathogens such as the microsporidium *Kneallhazia solenopsae*, a pathogenic fungus from South America that only infects *S. invicta* and *S. richteri* fire ants, making fire ant queens less fertile. Diverse communities of native ant neighbors that are quick to take advantage of fire ant colony disruption and bold enough to move in on their territory will also play a part in reducing the footprint of invasive ants.[22]

While wiping out invasive red fire ants completely is a long shot, biological control strategies such as these may help to bring *S. invicta* numbers down to more manageable levels, particularly in locations where the widespread distribution of fire ants and their proximity to crops and people makes it impractical to control populations with pesticides. In this rare instance of humans crossing paths with zombie-makers, the result is a partnership in which the zombifiers are on our side.

*

However, invasive fire ants aren't the only victims of phorid zombification. There's a species of phorid fly in the United States that specializes in parasitizing a very familiar and beneficial insect pollinator. Their stripey, fuzzy, buzzy target is currently facing threats on a number of fronts, among them climate change, pesticide use, parasites, pathogens, invasive predators, and habitat loss. Most people know this insect as the honey bee (*Apis mellifera*). But once it's been parasitized by the phorid fly *Apocephalus borealis*, it goes by another name: the "ZomBee."

This zombifying phorid implants its eggs inside honey bee hosts, and once the larvae hatch and take charge, the bee abandons its hive after dark and embarks on a grim nighttime journey toward the nearest source of bright light. It displays erratic behavior and gradually loses motor control until it finally topples over and dies, whereupon the phorid fly's larvae squirm out of its body to pupate, emerging from the junction between the bee's head and thorax.

Ironically, Americans' beloved *A. mellifera* is as much an invader in the US as are red imported fire ants. One of its common names is the European honey bee, and its native range is in Europe, Africa, and Asia. However, its introduction to the wider world began in the seventeenth century (European colonizers brought the bees to America in 1622), and the species is now established worldwide, in the wild and in human-managed hives. There are no native bee species in North America that live together in colonies as large as those of *A. mellifera*, so the collective efforts of honey bees—their "dances" communicate to thousands of fellow workers where flowers are located—can make them more effective pollinators than their North American bee cousins. As a result, the honey bee tends to be regarded as a helpful insect (from a human perspective) rather than an invasive and harmful pest (looking at you, red imported fire ants).

A. borealis phorid flies, unlike the honey bees, originated in North America and are known mostly from infected host insects found in the eastern and western regions of the US and Canada. Adult flies have yellowish bodies and measure 0.08 to 0.11 inches (2 to 2.9 millimeters) long—smaller than a fruit fly—and for decades they were known as parasites of yellowjacket wasps and several species of bumblebees. But in the early 2000s, when an entomologist in California went on an evening hunt for insects to feed a captive praying mantis, he unexpectedly discovered that *A. borealis* had branched out and was targeting a new victim—and leading it to display some very odd behaviors in the process.

In 2008, John Hafernik was a professor of biology at San Francisco State University (he's now professor emeritus), and he had just returned from an entomology field trip, where he'd found a praying mantis for his teaching lab. The hungry mantis needed its dinner, so Hafernik went looking for insects outside the campus's biology building, and he found some honey bees underneath the external light fixtures. The bees looked like they weren't doing very well—some were wandering in circles and some were lying on their backs twitching, with their straw-like tongues lolling out. But a meal is a meal, and so he scooped up some of the bees and brought them back to the lab, where the mantis eagerly devoured them.[23]

That spot became a daily destination for Hafernik to grab a quick takeout supper for his mantis, as there were always addled bees staggering around near the building lights, and they were relatively easy to capture. But then one day, when Hafernik returned to his office with the usual container of discombobulated bees, he set the vial down on his desk and forgot about it. When he came back to it, about a week later, there was a surprise waiting for him inside: the dead bees were surrounded by a bunch of small, brown pupae.

"I knew that the bees were parasitized, though I didn't really know exactly what was causing the parasitism," Hafernik says. He suspected the pupae belonged to phorid flies and were connected to the bees' peculiar "strandings" under the building's lights and inside the light fixtures. He reached out to phorid expert Brian Brown and started working with graduate students and other entomology colleagues to identify the parasite and get to the bottom of why the bees were behaving so bizarrely.

Morphological comparisons and DNA analysis confirmed that the phorids affecting honey bees were *A. borealis*, which had never been detected in honey bees before, the researchers reported in *PLOS ONE* (Brown was a coauthor).[24] In laboratory experiments, the scientists observed and documented how the phorids infect honey bees. A female fly flitted fleetingly over a honey bee before alighting on the bee's abdomen and inserting her ovipositor "for two to four seconds," the scientists wrote. Inside a bee, the growing larvae fed on their host's thoracic muscle tissue and possibly parts of the bee's nervous system. About seven days later, the larvae would emerge from inside their now-deceased bee host (see figure 11). On average, bees hosted up to 13 larvae; in one instance, 25 larvae wriggled out of a single bee to pupate. Adult flies popped from their pupae around 28 days after that.

As the larvae grew inside infected honey bees, the bees' behavior became stranger. Honey bees typically stay in their hives at night, but nighttime was when the scientists found parasitized bees crawling around near outdoor lights—even when the weather was so cold and rainy that no other insects were around.

"Stranded bees showed symptoms such as disorientation (walking in circles) and loss of equilibrium (unable to stand on legs)," the study authors wrote. Andrew Core, lead author of the *PLOS ONE* paper and a graduate student in Hafernik's lab at the

Figure 11 Phorid larvae exit the corpse of a "ZomBee." *Courtesy of John Hafernik*

time, said that the stranded, parasitized bees kept stretching their legs out, only to tip over and fall down. "It really painted a picture of something like a zombie," Core said in a 2012 interview.[25]

Or, rather, a "ZomBee," as the unfortunate host insects came to be called.

Most insects that are attracted to lights at night are highly active, but the stranded bees moved sluggishly, if they moved at all, and usually died the next day. Their colony abandonment and disorientation were similar to the symptoms exhibited by red imported fire ants parasitized by the behavior-manipulating phorid *Pseudacteon tricuspis*, the study authors noted. Molecular mechanisms for exactly how the phorids were zombifying their insect hosts—and how that might benefit the parasitoid—were unknown, but the scientists suspected that the flies may have been controlling the bees by altering their circadian rhythms or enhancing their sensitivity to light.

While the discovery of phorids' jump to honey bees was new, the presence of *A. borealis* in honey bee colonies was found to be widespread. When the researchers investigated honey bee hives in the San Francisco Bay Area, they detected signs of *A. borealis* in 24 of the 31 sample sites. In a honey bee hive that the scientists set up for observation near the San Francisco State University campus lab, the infection rate of honey bee workers reached 38 percent. One especially concerning detail was the presence of phorid pupae and empty pupal shells littering the bottom of the hive, along with a number of dead bees, "indicating emergence of adult phorids within the hive and the potential for *A. borealis* to multiply within a hive and infect a queen," the scientists wrote.

By adapting to parasitize honey bees, *A. borealis* may have unlocked an array of benefits that went beyond finding fresh bodies to zombify. Unlike bumblebees, which live in relatively small colonies that last for just one season, honey bees inhabit hives that may contain tens of thousands of individuals and are occupied and warm throughout the cold winter months. This creates a hospitable, long-term environment for the phorids and provides them with a constantly replenishing supply of potential ZomBees. Considering that commercial beekeepers in agricultural regions often maintain thousands of hives close together, a successful phorid invasion could quickly cascade into a population explosion with serious repercussions for the beehives they infest. Even if only foraging workers are parasitized, that could still dramatically affect the overall health of the hive, the researchers said. And the transport of infected commercial hives could introduce phorids far and wide, "given the number of states that commercial hives cross and are deployed in."

In fact, "the domestic honey bee is potentially *A. borealis*' ticket to global invasion," the study authors warned. "Establishment of *A. borealis* on other continents, where its lineage does not occur, where host bees are particularly naïve, and where further host shifts could take place, could have negative implications for worldwide agriculture and for biodiversity of non-North American wasps and bees." In the early 2000s, an emerging syndrome known as Colony Collapse Disorder (CCD), in which most of a colony's worker bees abandon the hive, leaving behind queen, food, and young, devastated commercial beehives and left beekeepers and scientists mystified as to the cause. Perhaps several dead workers might be found near an abandoned colony, but otherwise there were few clues to explain what went wrong. Viral disease, fungal parasites, and parasitic mites—or some combination of these—were considered to be the likeliest causes for hive abandonment. And now phorid parasitoids, the researchers suggested, should be added to that list, as a potential driver of CCD.

And if merely zombifying bees wasn't enough, phorids may also introduce other deadly pathogens into honey bee populations. The scientists found that phorid larvae and adults often carried a unicellular fungal parasite—*Nosema ceranae*, which causes Nosema disease (also known as nosemosis). Nosema disease causes premature death in honey bees and can spread quickly through a hive. Infected bees excrete spores that may contaminate a hive's food supply, and once those spores are in the hive, they may remain viable for months. Spores enter adult bees through their mouth and then germinate in the bee's digestive system, where they produce more spores. They spread as worker bees clean honeycombs soiled with spore-laden bee poop that accumulates inside the hive during the winter months.

Zombifying flies also carry a pathogen known as deformed wing virus, the *PLOS ONE* researchers reported. The virus causes bee larvae to develop with an array of physical abnormalities, including shriveled wings that are useless for flight.

To assess the current condition of how widespread *A. borealis*–zombified honey bees were in the United States, the scientists realized, would take a lot of eyes monitoring local bees. About six months after publishing the *PLOS ONE* study, Hafernik, along with colleagues at the Natural History Museum of Los Angeles County, and the Department of Biology and the Center for Computing for the Life Sciences at San Francisco State, introduced an online tool enabling citizen scientists to monitor honey bee zombies. They chose a name that quickly garnered media attention: ZomBee Watch.

The still-active project has three objectives: determine where *A. borealis* is parasitizing honey bees (originally North America, but now worldwide); track overall honey bee activity at night (whether the bees are parasitized or not, to gather data on light attraction after dark); and offer citizen scientists a chance to contribute to what is known about honey bees "and to become better observers of nature," according to the website (www.zombeewatch.org). The site includes instructions on how to build a light trap; how to collect and observe possible zombie honey bees or dead bees near lights that may contain *A. borealis* hitchhikers; and how to record and upload observations and other data, such as photos of emerging larvae, phorid pupae, or adult flies.

Reports from more than a thousand registered ZomBee Watch users, along with hundreds of samples, have helped map where honey bee zombification is happening. The scientists' early suspicions about the extent of the parasite's spread proved to be correct. Hafernik and his colleagues initially found evidence of zombie honey bees in Northern California and in South Dakota. Within a year after ZomBee Watch went live, reports were coming in documenting zombie bees "from Santa Barbara in the south all the way to Seattle in the north," Hafernik said in a statement for SF State News.[26] By 2016, *A. borealis*-parasitized honey bees had been identified in Oregon, Pennsylvania, Vermont, and Virginia in the US, as well as in Egypt and Canada. Nucleotide sequences—the building blocks of RNA and DNA—

matching those of *A. borealis* have turned up in screenings of honey bees from Belgium and South Korea, though no adult flies have yet been detected in those populations. And the reach of honey bee–zombifying flies has the potential to extend even farther, according to models produced by Hafernik and Brown, along with Sofia Croppi of the University of Bristol in the United Kingdom; and Chen-Yang Cai and Chao-Dong Zhu of the Chinese Academy of Sciences.

In 2021, Hafernik, Brown, Croppi, Cai, and Zhu evaluated the global invasion risk for *A. borealis* in honey bees, in the journal *Apidologie*. Based on reports submitted to ZomBee Watch, on studies about *A. borealis* sightings published after 1990, and on data from museum collections, the scientists mapped appearances of zombified bees in North America, identifying them in 123 sites in the US as far north as Alaska—and in Alberta, British Columbia, New Brunswick, Nova Scotia, Ontario, and Québec in Canada. They then built a model to predict the potential for the flies' global distribution. Phorids could easily hitchhike to new locations by tagging along inside infected bees, as honey bee hives are frequently transported across state lines and from country to country. And if climate conditions in their new home are hospitable enough, the flies could become established and seek out new honey bee hosts.[27]

Given the chance, opportunistic *A. borealis* could even target other native species as new hosts, the researchers warned. Prior to *A. borealis*'s discovery in honey bees, it was already known for parasitizing several species of wasps and bumblebees and had even been found in black widow spiders (*Latrodectus mactans*). In 2017, Brown reported the first known instance of *A. borealis* in carpenter bees, in Santa Barbara, California. A female valley carpenter bee (*Xylocopa varipuncta*), one of the largest native bee species in North America, was found behaving strangely, "buzzing and flying in circles on the ground, obviously in distress,"

Brown wrote. After the bee was collected, a total of 15 *A. borealis* larvae crawled from its corpse and pupated.[28]

According to the global invasion model, *A. borealis* would have little trouble adapting to places far from its native ecosystems in North America. It could likely find suitable habitats "across most of western, southern, and central Europe, parts of the Mediterranean Basin, Asia Minor, the Caucasus, extensive areas of central and southern Africa, eastern Asia, coastal areas of Australia, New Zealand, and coastal and mountainous regions of North and South America," the researchers reported. The zombifying phorid could also find welcoming conditions "on islands of the North Atlantic Ocean, Madagascar, New Caledonia, and the Aleutian Islands." Climate change could expand the footprint of hospitable ecosystems even more (though, that likely won't happen before the middle of this century). Measures such as quarantining bee shipments from high-risk regions for one to two weeks or restricting shipments to spring—the time of year when rates of phorid parasitism are seasonally at their lowest—may help to contain infestations and prevent the flies from becoming established in new locations.

As for understanding what makes a ZomBee, Hafernik—now retired—says there's still plenty of work to be done. There are many unanswered questions about the impact of phorid zombification on honey bee hive health in the long-term, and next to nothing is known about the mechanisms of how the zombifying phorids manipulate bees into flying at night and otherwise change honey bee behavior, "whether it's through gene expression or some other mode, or neurotransmitter-like substances," Hafernik says.

Another open question is if the phorids are manipulating ZomBees into leaving their hives, or if making that timely exit instead reflects a honey bee's typical reaction to infection. In 2022, researchers in Australia and Scotland reported that wounded or immune-compromised bees left their hives "seemingly by choice," perhaps recognizing that they posed a threat to the colony. In some cases, healthy honey bee workers would forcibly evict their ailing sisters, likely in response to chemical signals emitted by the sick bees.[29] However, more research would be required for scientists to determine if that's the case where phorid zombification is concerned.

"Is the fly manipulating the host?" Hafernik asks. "Or is the host simply responding to a disease, a parasite, a stressor, rather than actually being manipulated to take on a particular kind of behavior? We still haven't answered that."

The good news is that if you're an aspiring citizen scientist who wants to contribute to the nocturnal search for ZomBees, their night of the living dead is far from over.

9

WHAT LIES WITHIN

Zombifying Worms

"A zombie has no will of his own. You see them sometimes walking around blindly with dead eyes . . . not knowing what they do, not caring."

—GEOFF MONTGOMERY (RICHARD CARLSON), *The Ghost Breakers*, 1940

Avène in southern France has long been known for its water. Situated in a picturesque forested spot in the Parc Naturel Régional des Grands Causses, about 50 miles (80 kilometers) to the north of Montpellier, mineral-rich flows burble from a thermal spring deep in the wooded wilds. Since 1743, people have traveled there to bathe in the pristine waters, and centuries later, local spas and resorts still offer tourists the chance to experience the spring's alleged restorative powers. On a warm July night, a small group of water-seeking folk had gathered in Avène around a private, outdoor swimming pool, its surface glimmering in the darkness with the promise of a refreshing respite from the summer heat. But the visitors by the pool that evening weren't there for a spa retreat, nor were they anticipating a relaxing midnight swim. They were scientists waiting patiently for the arrival of other nighttime pool visitors: insects that typically spend their time on dry land and can't even swim very well but were nonetheless drawn irresistibly to the pool by the manipulation of a parasitic host. This host, a lengthy, threadlike creature known as a hairworm, was coiled snugly inside the insects' bodies and was marching them toward a watery doom.

In fact, by this final stage of the infection, most of a worm-carrying insect's body cavity—with the exception of its head and legs—was nearly filled up by the coiled parasite, which had grown to be many times the length of its host.

Scientists from universities in France and Germany had stationed themselves poolside in Avène during evenings over two summers, in 2000 and 2001. They kept watch over the water nearly every night, gathering evidence of a bizarre behavior in hairworm-parasitized insects; the behavior had previously been described anecdotally in scientific literature, with accounts dating as far back as 1922, but had not yet been formally studied. Hairworms were known to grow to adulthood inside the bodies of insects, but hairworm reproduction takes place in streams and rivers. Prior observations reported that when an adult hairworm inside a host was ready to reproduce, the host insect—a non-aquatic species—would trundle toward a nearby body of water and hurl itself in. The hairworm would then wriggle out of the insect's body and swim off to mate, leaving its former host to drown (or sometimes struggle back to land). Now, the researchers visiting the pool in France were conducting one of the first scientific investigations of this apparently suicidal insect behavior, to see if they could establish a definitive link to the influence and manipulation of a zombifying hitchhiker: a parasitic hairworm.

There are more than 350 described species of hairworms (also known as horsehair worms or Gordian worms) in the phylum Nematomorpha, though by some estimates there may be as many as 2,000 species distributed worldwide. Most of the known species are gordioids, which are semiaquatic and inhabit freshwater ecosystems on every continent except Antarctica. They are free-living as adults and parasitic as larvae, first seeking aquatic arthropod larvae as their intermediate hosts, then moving on to land-dwelling arthropods such as crickets, grasshoppers, mantids, and beetles. Inside these definitive hosts, the hairworms will grow to be mature adults. For more than a century after the first hairworm species was described in 1758 (*Gordius aquaticus*), they were thought to belong to a different group of worms called nematodes, or roundworms. But in 1885 they were classified as a separate phylum—the name "Nematomorpha" comes from the Greek word for "thread" (*nema*) and "forms" (*morphi*), referencing the threadlike appearance of their bodies and underscoring their resemblance to nematodes.

As their name implies, hairworms look like coarse strands of brown or black hair, measuring no more than 0.1 inches (3 millimeters) in diameter. Lengthwise, their bodies typically measure around 4 to 8 inches (10 to 20 centimeters) long on average, but some hairworms can grow to be up to 39 inches (1 meter) long. Now imagine all of that length coiled inside the hollowed-out abdomen of a cricket—and then spooling out from the cricket's soggy bottom, millimeter by millimeter, as the cricket flails helplessly in the water.

Hairworms mate in springtime, and describing their coupling as a knotty affair would be an understatement. In fact, they're known as Gordian worms because amorous adult worms gather by the dozens to form giant, knotted balls that resemble the Gordian knot of ancient Greek legend. According to the myth, the intricate knot was created by the Phrygian King Gordius and was so tangled it couldn't be undone—until Alexander the Great took his sword and slashed the snarled mass to bits. Once female hairworms emerge from these massive mating knots, they lay eggs in the water by the millions, in sticky, gelatinous strands that measure up to 24 inches (61 centimeters) long. The larvae hatch about a month later and are microscopic, their bodies no more than about 0.1 millimeters in length. But despite their small size, they are already on the lookout for an intermediate host to infect, typically targeting the aquatic larvae of flies and mosquitoes.

Hairworm larvae enter their hosts through the mouth or by using sharp structures around their own mouthparts to burrow through the host larvae's epithelium—the outermost layer of tissue surrounding the body. Once inside, the invader surrounds itself with a protective capsule-like cyst and hunkers down, passively waiting for the insect larva to become an adult fly, midge, or mosquito, and fly off to begin its life on land. If the larval hairworm hitchhiker is lucky (and its host insect is not), this intermediate host will be eaten by a terrestrial insect predator, perhaps a praying mantis, a beetle, or a cricket. On this next stage of the hairworm's journey, now in the body of a definitive host, the parasite emerges from its cyst and burrows through the new host's gut wall into the abdominal cavity. There, it feeds by absorbing nutrients through its skin, leaching them from the insect's tissues and blood, and it begins developing into its final form—a threadlike adult. Over the coming weeks, it will molt and grow.

And grow.

And grow some more.

Snug and secure inside an insect abdomen, hairworms feed on their host's hemolymph, fat deposits, and sex organs. And they get bigger and bigger until their coiled bodies fill up the host's entire abdomen; in 1940 an entomologist in Idaho found Mormon crickets (*Anabrus simplex*) with *Gordius robustus* hairworms snugly wrapped around the crickets' intestines.[1] Parasitic hairworms in mantids may grow to be so long and fill the host so tightly that their undulating movements visibly distort the surface of a host's abdomen (when I read about this, I was uncomfortably reminded of one of the weirder moments toward the end of my own pregnancy, when I could watch the contours of my kid shifting and stretching, as they kicked me in the kidneys). Most of the time, just one parasite will make its home inside an insect host. But hairworms are also shareworms, and scientists, in 1897 and 1967, respectively, reported finding two hairworms and four hairworms stuffed inside a single mantid's abdomen.

Parasitism can run rampant among an insect population in habitats where hairworms are plentiful. In northeastern Argentina's Misiones Province, mantids are frequently spotted spewing hairworms from their bottoms, in a grotesque parody of live birth.[2] This is such a common site that local Indigenous people referred to the mantids as *mboisy*—"mother of a snake" in the Mbyá language.

After about three months, a hairworm will need to leave its host in order to complete its life cycle. That exit, which never takes place through the host's mouth, needs to happen in the presence of water—preferably a natural body of water, but hairworm hosts will also discharge their zombifiers in livestock's watering troughs, pets' water dishes, open hot tubs, or even indoor toilets or aquariums, given the opportunity. Some accounts of the parasite's back door escape describe a hairworm squirming its way to freedom through the host insect's anus, though it's also possible that the worms wriggle out through exoskeleton gaps in the flexible layers connecting the hindmost segments of the host's posterior.

If you've never witnessed this firsthand, you're in luck: plenty of videos on YouTube offer a front-row seat to the entire spectacle.

One standout example is an episode of the PBS series *Deep Look*, titled "These Hairworms Eat a Cricket Alive and Control Its Mind," which includes footage captured by parasitologist Ben Hanelt, a principal lecturer at the University of New Mexico. Initially, the cricket struggles energetically as the worm begins to peek out of the insect's bottom. But then the cricket seems to resign itself to the indignity of its fate. The worm undulates in the water to drag itself out of the host, its body coiling in looping arabesques until its tail finally pops free and it swims away.[3] This procedure typically takes around ten minutes on average, depending on the length of the hairworm. But to ensure that the insect's rear end ends up in water, a mature hairworm must manipulate its terrestrial host into doing something it wouldn't ordinarily do: embark on what will very likely be a one-way quest for a creek, pond, or puddle.

Or, if the circumstance presents itself, a swimming pool.

*

Near the Avène pool, an expanse of concrete measuring about 16 feet (5 meters) wide separated the pool from the forest, creating a paved highway for parasitized, water-seeking insects and presenting the scientists with an unobstructed view of the zombified hosts on their procession toward a potentially fatal plunge. They tracked the insects visually "without disturbing them until they entered the swimming pool itself," the scientists reported in the *Journal of Experimental Biology*.[4] Once a hairworm had exited a drowning insect, the researchers collected both host and parasite, preserving them in alcohol. The researchers captured some of the parasitized insects alive before they expelled their hairworm zombifiers, for closer examination later in the lab; they also compared pool-bound insects to insects that they collected in the forest.

Over two summers, the scientists observed this seemingly suicidal pool-plunge—followed by the appearance of a squirming worm—in nine insect species and two spider species. Parasitizing the insects were two species of hairworm: *Paragordius tricuspidatus* and *Spinochordodes tellinii*. Nearly every insect that approached the pool eventually entered the water. Once an insect dove in, a threadlike hitchhiker began to peek out from the host's bottom—sometimes spooling out immediately and sometimes unwinding over several minutes and swimming away after the insect had drowned. If the researchers removed an insect from the water before it had discharged the parasite, the zombified arthropod promptly returned to the pool and jumped in again.

The most common species to visit the pool during the month of July was the flightless wood cricket (*Nemobius sylvestris*). To find out if the behavior of infected crickets differed from the behavior of those that didn't carry a parasitic stowaway, the researchers first organized a cricket sleepover. They collected 38 *N. sylvestris* crickets from near the edge of the swimming pool, and 33 individuals from the forest, at a distance of at least 328 feet (100 meters) from the pool, and kept the insects overnight in a terrarium. The next night, they brought the insects to the pool and placed them on the concrete area at a distance of about 7 feet (2 meters) from the water's edge in groups of four—two crickets from the forest, and two that had been collected poolside. The scientists then monitored the insects for the next 15 minutes. "When a cricket entered the water, the experiment was completed for this individual," the researchers said.

After the experiments were done, all the insects were brought back to the lab for euthanization and dissection, to see if they contained a hairworm. Of the crickets that had originally approached the pool, 95 percent were infected (compared to just 15 percent of the crickets taken from the forest). "Among these 41 infected insects, 36 harboured one worm, four harboured two worms and one individual harboured four worms," the scientists wrote.

In the next round of experiments, they again collected crickets from the forest and from the side of the pool. They tested the crickets' reactions to water in a Y-shaped maze. Both of the maze's branching arms ended in troughs: one dry, one filled with water. Pumps in the maze's arms wafted air from each tunnel back toward the crickets, "to increase the possibility of water detection," according to the study. Crickets were given 30 minutes to explore the maze and choose a branch—humid or dry—and were then removed and dissected.

None of the crickets taken from the forest were infected. But of the crickets that were collected poolside, 94 percent carried hairworms. In the maze, both uninfected and infected crickets visited the branch that held water. But once they were in that branch, infected crickets were more likely to jump in the water. The parasite's presence seemed to change their behavior, causing them to do something that was potentially deadly to the host but beneficial to the parasite.

However, the hairworm's exit strategy isn't risk-free. By the time a hairworm drives a host to water, it has grown so long that it can take up to 28 minutes to wriggle out. A cricket struggling in water is a tasty-looking target for fish and frogs (not to mention the added attraction of a wriggling worm hanging from its butt), and scientists from Denmark and France wondered if a hairworm could survive being swallowed by a predator. The researchers captured infected crickets (also at Avène) and then observed 477 lab-supervised predation attempts in aquarium tanks. Green paddy frogs (*Hylarana erythraea*), largemouth bass (*Micropterus salmoides*), pumpkinseed sunfish (*Lepomis gibbosus*), and rainbow trout (*Oncorhynchus mykiss*)—common predators in freshwater ecosystems in France—all snapped up *P. tricuspidatus*-infected wood crickets after the crickets entered the water. Astonishingly, more than 100 hairworms performed daring escapes after they and their host had been swallowed, slithering out through a predator's mouth, nose, or gills. (Alas, their cricket hosts weren't so lucky.)

"To our knowledge, this escape response by a gordian worm is the first example of a parasite or any organism surviving predation in this way," the scientists wrote in *Nature*. If the worm hadn't reappeared after five minutes, it was presumed dead, having succumbed to "the hostile environment of the predator's stomach."[5]

But how do the hairworms control their insect puppets and direct them toward water? One possibility is that a worm could somehow enhance its host's attraction to moisture. However, the Y-shaped maze experiments seemed to contradict that idea—infected crickets chose the dry branch of the maze about as often as they picked the branch with water in it—though, once they found the water, they would usually dive right in.

One piece of the manipulation puzzle was uncovered a few years later by scientists from France and the United Kingdom. These researchers investigated a type of bright-green grasshopper in the species *Meconema thalassinum*—also known as drumming katydids—that were infected by the hairworm *Spinochordodes tellinii*. Perhaps, the researchers suggested, answers might be found by looking at proteins produced by cells of infected grasshoppers and their hairworm parasites during three stages of behavior manipulation—"before, during and after the host jumps into water," the scientists wrote in *Proceedings of the Royal Society B*.[6]

So they looked at proteomes—overviews of all of the proteins produced by an organism's cells and tissues—in the host insects and their parasites. In a zombified *M. thalassinum* grasshopper, certain proteins associated with neurotransmitter activities went zinging through its brain during the insect's approach to water and then again when it made the fatal leap. As for the hairworm hitchhiker, its proteome also changed. It, too, produced a special suite of proteins prior to and during the grasshopper's approach and jump into water. Novel proteins were also present after it emerged from the rear end of the host.

When *S. tellinii* was manipulating the insect's water-seeking attempts, the parasite produced two families of proteins that prior studies linked to the release of neurotransmitters. Another group of proteins released by the parasite at that time is connected to regulation of a self-destruct mechanism in cells, called apoptosis. Other types of parasites are known to change host behavior through a similar path. By causing cells in a host's central nervous system (CNS) to self-destruct, these proteins disrupt normal function and change chemical signals in the brain. This disruption of the CNS leads to "aberrant behavioural responses" in the host, the researchers reported.

As for the mechanisms that might attract an infected insect toward water, another research team from Japan and Taiwan pondered this riddle and saw the light—literally—finding that hairworm-infected mantids were fatally attracted to light reflecting off water.

Giant Asian mantises (*Hierodula patellifera*) are often parasitized by hairworms in the genus *Chordodes*. Prior research hinted that *Chordodes* manipulation could boost a host's affinity for light. But brightly illuminated spots can reflect off many surfaces in the mantids' habitats and don't necessarily promise the presence of water. However, the researchers pondered, what if the parasite's manipulation sharpened its host's attraction to a specific quality of light—not intensity, but polarization? When sunlight bounces off water, the light is horizontally polarized, meaning that the direction of the light's energy waves oscillate parallel to the surface. Perhaps some neural tweaking by the hairworm host made this type of light uniquely attractive to parasitized mantids, and compelled them to seek water and take the plunge.

To test that, the scientists placed infected mantids in a cylinder with one end illuminated by polarized light and the other lit up by unpolarized light. They found that more of the mantids hopped toward the polarized light. But they only did so if it was horizontally oriented, like light on a liquid surface. Next, the researchers set up outdoor experiments in a mesh enclosure at Kobe University. Inside the enclosed area were two pools: one reflecting light that was brighter than the other but was weakly polarized and the other reflecting dimmer light that was strongly polarized. Both lights were horizontally polarized. The researchers then released infected mantids into a tree near the two pools and waited to see if the insects would approach the water and which pool the parasitized puppets would choose.

Of the 16 infected mantids that approached and entered the water, 14 chose the pool that reflected strongly polarized light, suggesting that the hairworms' manipulation of their zombie hosts had something to do with the mantids' perception of the reflected light's horizontal polarization.

"To our knowledge, this is the first study demonstrating that a manipulative parasite can take advantage of its hosts' ability to perceive polarized light stimuli to alter host behaviour," the scientists wrote in *Current Biology*. "Animals have evolved a great diversity of photosensitive visual systems, which allow them to detect intensity, colour, contrast and polarization of light. In the course of their co-evolution with those animals, parasites may also have evolved to take advantage of the diverse visual systems of their insect hosts so as to enhance host manipulations."[7]

For an infected mantid to end it all by going towards the light sounds almost poetic. But is that fateful dive truly the end for a zombified host? As it happens, there's a faint glimmer of light at the end of the zombie tunnel for hairworm hosts—that final leap into water that rids them of their pesky passenger isn't necessarily a death sentence. In fact, if the insect doesn't drown, it can live for several months, researchers from Australia, Switzerland, Canada, and France reported in *Behavioral Ecology*.[8] The scientists found that *N. sylvestris* crickets parasitized by *P. tricuspidatus* hairworms were attracted to light when the hairworm was ready to emerge, which ultimately led them to water. As this attraction was part of the parasite's manipulation, the researchers predicted that once the parasite was gone, the insect would no longer be unnaturally attracted to light. They collected infected crickets in Avène from the concrete areas on the forest outskirts near water—about 99 percent of the crickets in these locations contained hairworms—and gathered uninfected crickets from deeper within the woods.

In the experiments, all infected crickets walked toward a bright cone of light, compared with just 52 percent of the uninfected crickets. Zombified crickets would also walk farther and faster to get to the light and took a more direct route to get there. Ex-zombies that had been abandoned by their hairworms were still attracted to light—at first. But that attraction started to fade once the zombifier was gone. After about 20 hours, it was completely snuffed out, and the crickets showed no more interest in light than uninfected crickets did.

However, even after 35 hours, the walking patterns in ex-zombies hadn't yet recalibrated to normal; they still walked faster and longer than uninfected crickets did. This hinted that hairworms dose their host's central nervous system with a chemical cocktail containing multiple compounds—some that affect light sensitivity and some that affect walking patterns. Select compounds may continue to manipulate behavior even after the parasite is gone. And even if the insect survives, those changes may be irreversible.

"Given that crickets can live up to 130 days after the emergence of their parasite," the researchers wrote, "further studies would be necessary to precisely determine how the different postinfection alterations are maintained."

*

If organisms' success were solely judged by how numerous they are, tiny and threadlike worms called nematodes would be crowned the most successful creatures on the planet. The phylum Nematoda includes nearly 30,000 described species, though the number of species in existence is thought to be much higher, perhaps a million or more. Also known as roundworms, nematodes have tubular bodies in which a pressurized, fluid-filled inner cavity is surrounded by muscle and protected by a tough outer shell called a cuticle. Their bodies are tapered, with a mouth at one end and an anus at the other, and they can be found everywhere on Earth. Nematodes may live in soil, polar ice, or thermal springs; in freshwater or saltwater; or as parasites of plants or animals (including other nematodes). Subterranean land species have even been discovered living 5.6 miles (9 kilometers) underground, and some marine species dwell at depths more than 29,000 feet (9,000 meters) below sea level. There are fossil nematodes dating to around 400 million years ago, but the first nematodes are thought to have evolved approximately one billion years ago.

The smallest nematodes are microscopic, with many species measuring less than 0.04 inches (1 millimeter) long. But the biggest known nematode—*Placentonema gigantissima*, which parasitizes the reproductive organs of female sperm whales—can measure a whopping 26 feet (8 meters) long. Individually, many of Earth's nematodes are too small to be seen with the naked eye. Collectively, however, is another story; their global abundance is nothing short of remarkable. Scientists have reported finding more than 200 nematode species in just one cubic centimeter of soil, and several million nematodes may comfortably inhabit a square yard of agricultural or woodland habitat. In fact, nematodes are so plentiful that there may be nearly 60 billion of these tiny creatures for every living human; by some estimates, if you were to count heads of every living animal on Earth, approximately four out of five of them would be nematodes.

One of nematodes' most ardent fans was the nineteenth-century biologist Nathan Augustus Cobb, renowned for his pioneering work on nematodes and hailed as the "father of nematology" in the United States. Cobb famously wrote that if all the nematodes living in the soil of a ten-acre field were arranged in a single-file line, they would form a parade that would stretch all the way around the planet. Cobb spent much of his career thinking about nematodes, and he used poetic and occasionally eerie terms to embrace the scope of their global abundance. He proposed that if all matter in existence—except nematodes—suddenly disappeared, a spectral afterimage would remain of Earth's vanished life, landscapes, and civilizations.

Their outlines, he imagined, would be sketched in the lingering presence of the billions upon billions of nematodes that were left behind.

"Our world would still be dimly recognisable, and if, as disembodied spirits, we could then investigate it we should find its mountains, hills, vales, rivers, lakes, and oceans represented by a film of nematodes," Cobb wrote. "Trees would still stand in ghostly rows representing our streets and highways. The location of the various plants and animals would still be decipherable, and, had we sufficient knowledge, in many cases even their species could be determined by an examination of their erstwhile nematode parasites."[9]

Cobb's infatuation with nematodes might have been deeper and more profound than most scientists', but he wasn't the first to find these tiny worms absolutely fascinating. People have observed nematodes for thousands of years, with descriptions appearing in medical literature in China as early as 2700 BCE, and in the ancient Egyptian medical text the Ebers Papyrus, dating to around 1500 BCE. In terrestrial soils and in marine environments, free-living nematodes are decomposers, and they play an important part in breaking down plant and animal matter, bringing oxygen into deep sediments and maintaining ecosystem health. However, approximately 60 percent of nematode species are parasitic, infecting plants and animals—including about one billion humans worldwide. And in some instances (fortunately not in humans), those parasitic infections come with a side order of mind control, with behavior manipulation and body distortion that increase the likelihood of the host being eaten by a predator so that the nematode can reproduce.

Such is the case for the arboreal ant *Cephalotes atratus*, also known as the common giant turtle ant. This burly canopy-dweller lives among the trees in tropical forests in South and Central America. With a dark black body measuring up to 0.8 inches (20 millimeters) long, a large, flattened head decorated with spikes and flanges, and mandibles powerful enough to chew through wood, the ant is an imposing sight. "The hard spiny armor is sufficient to protect it from any ordinary enemy," American entomologist William Montana Mann wrote about *C. atratus* in 1916.[10] But the tiny nematode *Myrmeconema neotropicum* is no ordinary enemy, and all of the giant turtle ant's impressive adornments can't keep it safe from this insidious parasite.

M. neotropicum lays masses of eggs inside the ant's gaster—the bulbous rear part of the abdomen—triggering a transformation that makes the ant's bottom resemble the ripe, red berries produced by many local plants. The berries are a favored snack of fruit-eating birds. The parasite also makes the ants slower, less feisty when threatened, and prone to waggle their rears in the air. This butt-wagging behavior is known as gaster flagging and likely draws the attention of birds with a taste for brightly colored berries.

Entomologists from the United States first glimpsed the peculiar red-bottomed ants in 2005 during field work on Barro Colorado Island, in Gatun Lake in Panama, where they were investigating gliding behavior in *C. atratus*. A few years later, the researchers published their findings about the berry-butted insects in the *American Naturalist*. It was "the first documentation of parasites causing apparent fruit mimicry in an animal host to complete their life cycle," the scientists reported.[11]

One of the study authors, insect ecologist Stephen Yanoviak, a professor at the University of Louisville in Kentucky, says that he and his colleagues were puzzled by the sight of the red-bottomed ants, which were foraging alongside normal *C. atratus* workers. They initially assumed that the different-colored ants belonged to different species, so it was odd to see them foraging peacefully side-by-side. Other scientists had previously suggested that the vivid red hue was simply a color variation within the genus. But when Yanoviak dissected some of the red-gastered ants and peered at the results through a microscope, he discovered that their gasters were stuffed with hundreds of eggs.

"I brought them into the lab and I broke open the red rear end of the ant, and out poured all these nematode eggs," Yanoviak says.

And inside each egg was a small, coiled worm.

Further examination showed that nematode infection reduced pigment density in the ants' rear ends, turning them from dark black to translucent amber. The color of the densely packed and yellowish nematode eggs was then visible through the ant's exoskeleton, turning its abdomen bright red. When the scientists measured how light of different wavelengths and across spectra reflected off the discolored gasters, they found that the light's properties resembled those of light reflecting off ripe red berries of the neotropical plant *Hyeronima alchorneoides*.

In other words, a bird searching for berries could easily be drawn to the sight of the ants' carmine cabooses and mistake them for fruit.

The redness of the ants' gasters increased as the parasites developed from embryos into first-instar worms. Infected ants were also about 10 percent smaller than uninfected ants, possibly due to the metabolic cost of hosting the parasites, which infected the ants in their colonies' nurseries as larvae. However, infected adults were also around 40 percent heavier than their uninfected nestmates, due to the load of parasites they were carrying.

The ants' overloaded abdomens were extremely fragile and prone to breakage, the scientists found.

"Those red rear ends would fall off of them if you handled them with too much force, which is not normal for these ants—these ants are like little tanks," Yanoviak says. All that extra parasite weight seemed to weaken the exoskeleton junction where the gasters attached to the rest of the abdomen, and the gasters could easily be plucked from infected ants' bodies—something that wasn't possible to do with healthy insects.

Infected ants also displayed unusual behaviors that seemingly benefitted the parasites, by increasing the likelihood of a fruit-eating bird targeting a berry-like ant. Ants with the reddest gasters—and the highest concentration of parasites—tended to have gaits that were unnaturally erect and unstable. They were also generally sluggish and less aggressive; they didn't bite when they were handled and didn't release alarm pheromones, which produce an unpleasant odor that smells like camphor and is pungent enough for humans to smell. But what really stood out was their gaster-flagging. When ants with parasites in their bright-red bottoms were outside their nests and hunting for food, they would hold their gasters up "almost continuously," the study authors reported.

"Presumably it makes them more conspicuous," Yanoviak says. "Some ants will do this to disperse defensive chemicals in the air, or to disperse their pheromones to warn other ants of a potential danger. But these ants with the parasites in there just do it all the time. The whole package suggests that this parasite is causing not only a morphological change but also behavioral and physiological," he explains. "It's pretty remarkable."

While the scientists didn't directly observe birds preying on the parasitized ants, they tested birds' attraction to red gasters by crafting wee, gaster-size balls of clay in different colors. They found that birds pecked at red and pink clay balls more often than all the other colors combined. The researchers also fed infected ants to chickens and examined the hens' feces for traces of nematodes, to confirm that birds could play a role in parasite reproduction. Sure enough, after a chicken was fed an infected ant, within 13 hours, hundreds of nematode eggs showed up in its excrement. And foraging *C. atratus* ants frequently pick up bird feces and carry it back to their colonies—nearly 70 percent of the solid material that they transport is bird poop.

M. neotropicum nematodes mature and reproduce inside ants, but how do they get inside their hosts in the first place? That's where birds enter the picture—rather, their poop does. Insect-eating birds typically ignore the giant turtle ants because of their spiky armor and pungent pheromone defense. But vividly red, nematode-stuffed gasters of infected ants look like a delicious treat to fruit-eating birds—and are highly visible if the tree-climbing ants are holding them up invitingly. The scientists hypothesized that birds would pluck and devour the gasters and then poop out the nematode eggs high in the treetops. *C. atratus* ants would then collect the parasite-laden poop, bring it back to their nests, and feed it to larvae. Inside the larvae, nematodes developed into adults. Eventually, they migrated into the host's gaster to mate and lay eggs, transforming the ant's black gaster into one that was bright red. Around this time in the ant's life cycle, it becomes an active forager, spending more of its time outside the nest. Fruit-eating birds snapped up the forager's berry-mimicking bottom, "and the parasitic life cycle repeats," the study authors wrote.

Scientists still have questions about how exactly the ants become infected. "We don't know if the nematode eggs hatch inside the colony and then burrow into the walls of the ant larvae, or if the ant larvae ingest them as eggs," Yanoviak says. "We would have to maintain some colonies of the ants in the lab and then experimentally infect them." One thing is certain: however *M. neotropicum* infects giant turtle ants, it's very, very good at it. *C. atratus* lives in forests from Panama to Argentina, and wherever colonies of these ants are found, the nematode zombifier is there, too. The coevolution of this fruitful host-parasite relationship "illustrates the diversity of amazing interactions that exist among species on this planet, and further illustrates how little we really know about what's going on out there," Yanoviak added. "It shows us how complicated and intricate the natural world is."

*

Berry-butted ants are the only known example—so far—of a parasitic manipulator making its zombie host look like a piece of fruit. However, zombifying worms have other colorful tricks up their sleeves for physically changing their hosts appearance as well as their behavior. One of the flashiest of these transformations is the handiwork of *Leucochloridium*, a genus in the phylum Platyhelminthes, or flatworms. Flatworms have a body plan similar to that of nematodes. But as their name implies, their forms are nearly two-dimensional. These worms don't have a respiratory system, so a flattened shape ensures that oxygen can diffuse through as much of their body's surface area as possible. Their only internal cavity is a branching gut, which ferries nutrients around their insides, and a single opening at one end serves as both mouth and anus, taking in food and expelling waste. Some types of flatworms are free-living, but the class known as trematodes, or flukes, are parasitic. There are about 9,000 known species of trematodes, and as adults, most trematodes parasitize animals with backbones—including humans.

Leucochloridium flatworms are a type of trematode, and they reproduce in the intestines of birds. But at least ten species are known to reach the guts of their definitive bird hosts by hitching a ride inside snails, mostly in the family Succineidae, or amber snails. Snails pick up the parasite when they eat the flatworm's eggs, which are excreted in the feces of infected birds. The eggs then hatch in the snail's gut, and the flatworm larvae migrate into its tissues. Immature flatworms will eventually induce infected snails to wander from the safety of leafy undergrowth and climb upward into the open, where they're more easily spotted by hungry birds searching for prey. And to make sure that insect-eating avians target the snails, the flatworm larvae perform an eye-catching act of insect mimicry: they produce pulsing, striped broodsacs that are stuffed with up to 200 larvae called cercariae, in jelly-covered cysts. These sausage-shaped broodsacs migrate into the snail's translucent upper tentacles—their eyestalks (see figure 12). As the bulbous, patterned broodsacs swell and stretch the mollusk's eyestalks, their rippling undulations—at a rate of about 60 to 80 contractions per minute—resemble the mesmerizing pulsations of 1960s-era psychedelic light shows (or the pupils of cartoon characters under the influence of a powerful mind-altering drug).

More importantly, from the perspective of an insect-eating bird such as a crow, finch, sparrow, or jay, these displays inside the snail's engorged eyestalks look like the movements of crawling caterpillars. Only, these tempting "caterpillar" treats contain hundreds of trematode larvae. Once swallowed by a bird, they settle down in their new home: the bird's cloaca, the all-purpose opening in its rear for urination, excretion, reproduction, and egg-laying.

Colors and band patterns of the flatworms' broodsacs vary by species. For example, *Leucochloridium passeri* sacs have a brownish tip, followed by light brown bands, greenish bands and

Figure 12 *Leucochloridium* parasitizing a snail in the family Succineidae. Each of the snail's eyestalks is stuffed with a parasitic brood sac. *Courtesy of Gilles San Martin*

patches, and then a wide brown band. Color bands in *L. paradoxum* broodsacs are mostly shades of green and yellow, with some black spots. Some infected snails will have just one distended and throbbing eyestalk. But flatworms typically produce multiple broodsacs inside their hosts, pushing these sacs into both of the snail's tentacles. And if a snail becomes infected by two species of flatworm, the mollusk may end up with swollen, palpitating eyestalks that display two different-colored broodsacs, with dueling parasitic displays pulsing their uniquely colored, come-hither signals: one bulging tentacle green-banded and the other in shades of brown.

A sac growing inside a snail takes about four weeks to become visible from the outside, and broodsacs typically reach their full development another four weeks after that, German physician and zoologist Ernst Zeller reported in 1874. By the time a broodsac attained its full size, Zeller noted, the membrane of the snail's tentacle was stretched so thinly around it that even slight pressure from outside the eyestalk was enough to rupture it and release the sac within. If that happened, the broodsac remained attached to the snail by a small structure at the sac's bottom, though it could be easily removed from the snail without the host appearing to suffer from the separation. After its removal, the broodsac would continue to pulsate, Zeller wrote. And under favorable conditions—for the parasite, that is—another brood sac might develop in the snail to replace the previous one.[12]

Zeller also directly observed birds' attraction to the parasitic broodsac's caterpillar-like appearance, according to a summary of his research that was published in 1875 in the *Annals and Magazine of Natural History*. When Zeller presented a tame warbler with an infected snail sporting a broodsac-stuffed eyestalk, the bird "immediately came down upon one of these sacs, tore it out of the tentacle, and swallowed it." Tests with other birds produced the same outcome. "In all these experiments it was observable that the bird, after having seized the *Leucochloridium* and torn it out with a single strike of the bill, swallowed it, sometimes immediately, sometimes only after striking it several times against the floor of its cage or the perch, thus behaving exactly as the insectivorous birds do with their ordinary food."[13]

In 1922 and 1931, other researchers suggested that *Leucochloridium* parasites not only changed what the snails looked like; they also affected the snails' behavior, prompting them to emerge from the gloom of leafy cover and perch on the sun-facing side of leaves, where their pulsing, caterpillar-mimicking eyestalks would be exposed and highly visible to birds. However, those conclusions were based solely on observations of infected snails and hadn't been formally tested through comparisons to snails that were parasite-free. Nearly a century later, that question caught the attention of biologists Wanda Wesołowska and Tomasz Wesołowski at Poland's Wrocław University. They decided to find out if *Leucochloridium* truly did manipulate how the snails behaved, by observing and comparing the habits of parasitized hosts to those of healthy snails.

Over seven days in May and four days in June, the scientists hunkered down in Białowieża National Park, an old-growth forest in eastern Poland where *Leucochloridium* trematodes were known to infect snails. In the forest, alder, ash, and spruce trees spread their branches to form a canopy overhead. Below, on the forest floor, grows a dense, nearly unbroken cover of meadowsweet, nettles, thistle, and reeds, dotted here and there with the vivid yellow of irises and marigolds.

"We slowly moved through the area, looking for snails with pulsating broodsacs," the researchers wrote in the *Journal of Zoology*.[14] They identified 30 infected snails, as well as 90 snails that didn't have visible broodsacs. All of the snails were European amber snails (*Succinea putris*), and the flatworms all belonged to the trematode species *L. paradoxum*. Wesołowska and Wesołowski then compared activity between the two snail groups, noting the density of leafy cover as well as the height and brightness in the locations where the snails crawled.

Though there were plenty of snails in the study area, the scientists found infected snails only in glades that were brightly lit. Compared with snails that didn't have visibly pulsating broodsacs, flatworm-hosting snails behaved in ways that seemed more likely to draw a predator's attention, putting themselves in danger but helping the parasite reach its definitive bird hosts. Infected snails moved around more than snails without broodsacs, and they spent more time in sunny spots where they were exposed and vulnerable to predatory attacks. They also favored locations that were higher up in the forest vegetation and tended to migrate toward the upper side of exposed leaves.

"All these changes in the snail behaviour would be beneficial for the parasite," the scientists wrote. Sight-hunting birds could easily spot active snails in such exposed locations, and could swoop down to snap them up without having to land on the forest floor. "The observed behavioural changes fulfil [*sic*] the criteria of behavioural manipulation," Wesołowska and Wesołowski wrote. "It seems justifiable to claim that, additionally to its own phenotypic modifications (production of colourful pulsating broodsacs), *L. paradoxum* manipulates also the behaviour of its intermediate *S. putris* host."

It's yet to be discovered exactly how *Leucochloridium* compels snail zombies to behave this way. But another species of trematode—*Dicrocoelium dendriticum*, also known as the lancet liver fluke—has provided scientists with some clues about how flatworms achieve zombification.

D. dendriticum flatworms, which can measure up to 0.4 inches (10 millimeters) long, are found across Africa, Asia, and Europe, and have been spotted sporadically in North America. Sexually mature adults lay their eggs in the gallbladder and bile ducts of ruminants, such as cows, sheep, or goats (the flatworms can also breed in other animal hosts, including humans). However, prior to the egg-laying stage of their lives, the worms require not one, but two successive intermediate hosts: land snails and ants. While snails play a necessary role, they do not seem to be manipulated by the parasites. Rather, *D. dendriticum* reserves zombification for its ant host, with one of the many larvae inside the insect's body departing from the rest of the group to invade the ant's brain. If a song gets stuck in your head on endless repeat, you might complain that you're the victim of an earworm. But ants infected by *D. dendriticum* experience the presence of an actual worm inside their heads—a *hirnwurm* (German for "brainworm").[15]

Getting into the ant's brain is a lengthy trip for the parasite. Its eggs are excreted by grazing mammals—the worm's definitive hosts. Snails eat the eggs, which hatch in the snails' intestines as miracidia, a type of larvae that are covered with tiny hairs called cilia. Miracidia burrow through the snail's gut wall into vascular tissue; there, they become sporocysts. The sporocysts then travel to the snail's digestive gland and, through asexual reproduction, yield another larval stage: free-swimming cercariae. Approximately three months after the snail swallowed the flatworm's eggs, it expels hundreds of cercariae inside slime balls. These are eagerly devoured by ants, and several dozen larvae may take up residence inside a single ant host. Inside an ant's abdomen, cercariae lose their tails and transform into another larval stage: metacercariae, which infect and then reproduce in ruminants.

Once *D. dendriticum* is inside an ant, its journey is nearly complete. But it still needs to find its way back to a ruminant, to lay its eggs. Grass-eaters typically don't eat ants, but if an ant happens to be attached to greenery where a ruminant is feeding, the herbivore will unknowingly swallow the insect along with its plant dinner. *D. dendriticum* must therefore seize control of its host's brain and compel the insect to climb a piece of vegetation where it's likely to be eaten, so that the trematode larvae can then escape into the grazer's body. Most of the larvae in an infected ant hang out in protective cysts, in the ant's gaster. But one or two *hirnwurms* migrate to the ant's suboesophageal ganglion—a part of the brain that controls the insect's mouthparts—where their zombifying work begins.

An ant harboring *D. dendriticum* will behave normally for most of the day. But when night falls and temperatures drop, the brainworm takes over. Up goes the ant, climbing to the tip of a flower or blade of grass, where it stops and bites down. Should air temperatures remain cool, an ant zombie will remain in the same spot for up to seven days, and it won't eat or defend itself while it's there. If temperatures warm to about 64 to 68 degrees Fahrenheit (18 to 20 degrees Celsius), a perching ant will release its grip and rejoin the colony. But that reprieve is only temporary. Obedient zombie ants return to their elevated posts night after night—until they're eventually eaten. This type of zombification is unusual because it seems to wax and wane; the ant behaves abnormally under certain temperature conditions, but in the absence of those conditions, it reverts to its normal behavior. Scientists were curious about the mechanisms behind this manipulation, and so a team of biologists and imaging experts peered inside infected ants' heads in an attempt to solve that mystery.

Researchers from the United Kingdom, Spain, and Canada used micro-computed tomography (micro-CT) scans to visualize trematode-infected ants' insides in three dimensions. They collected *Formica aserva* ants at Cypress Hills Interprovincial Park in Alberta, Canada, from a site where ants were known to be infected with *D. dendriticum*, gathering 20 zombified ants that were clinging to flowers, and 20 uninfected ants. The researchers then prepared the individuals for scanning. Images of infected ants' tiny bodies—workers measure no more than 0.4 inches (10 millimeters)—revealed that their gasters held at least 6 trematode larvae that were passively resting inside cysts, and one ant contained 98 encysted trematodes. Scar-like marks on the ants' gut walls suggested that the larvae had burrowed through the walls and into the gasters after they were eaten.

Micro-CT scans of the ant heads told a slightly different story: these *D. dendriticum* larvae weren't passively waiting inside cysts. Rather, they were actively piloting the ant, with CT scans delivering the first evidence that the parasites were physically touching the ants' brains. All of the ants that were collected while clinging to flowers had at least one *D. dendriticum* brainworm. One scan showed a brainworm with its oral sucker "in direct physical contact" with the front part of the ant's suboesophageal ganglion, and another infected ant had three worms in its head—though, only one of the parasites was directly touching the ant's brain tissue.

By invading the ant's head and touching this part of its brain, larvae gain access to neurons controlling locomotion and mandible muscles and can thereby manipulate ant movement as well as its attachment to plants. The scientists' detailed images hinted that future studies could use micro-CT scanning and 3D digital models built from the scans to explore and even unlock the mechanisms behind "the parasite's practice of on-and-off puppeteering," the researchers wrote in *Scientific Reports*.[16]

*

From brain-invading worms to brain-stabbing wasps; from body-melting bacteria to fruiting fungi—spend enough time with these zombifying parasites and their hapless victims, and it may seem like the zombie horror of the natural world is never-ending. Every unwilling partnership tells a gruesome story of body distortion and modified behavior, with each example more horrific than the last. But after years of studying the chemistry and neuroscience that underlies the relationships between zombies and their zombifiers, scientists have made numerous discoveries that benefit the living. Some of their findings illuminate a road toward the development of novel medicines for humans, including treatments for infections, autoimmune diseases, and cancers. Other discoveries about zombification offer possibilities for controlling invasive insect populations that threaten agriculture. And with popular interest in zombies spreading faster than a fictional zombie plague, the real science behind zombie insects, arachnids, mollusks and other invertebrate life lends insights into zombification as a pathway to broader lessons and research about evolution, biology, social infrastructure, and how we interact with each other. As it happens, zombies might have a thing or two to teach us about being better humans.

10

OUR ZOMBIE FUTURE

"You're all going to die down here."
—THE RED QUEEN (MICHAELA DICKER), *Resident Evil*, 2002

At this point you might be thinking (or possibly hoping) that mammals—and by extension, humans—are safe from real-world zombifiers. If only that were the case. It's certainly true that we're safer from zombification than arthropods. While the zombie manipulators that you've met thus far in these pages—from fungi and viruses, to wasps and worms (along with many, many others that aren't described here)—that target bugs are far more numerous than the organisms that can mind-control mammals, that doesn't mean that mammals are immune to having their brain chemistry hijacked by parasites. Certain microbial zombifiers have well-established track records for altering behavior in mammals, and humans are no exception.

Take rabies, for example. Caused by the rabies virus (RABV) in the *Lyssavirus* genus, rabies is a deadly disease found only in mammals. It targets the central nervous system, leading to brain inflammation, altered behavior, muscle paralysis, and death. Any mammal can be a carrier, but canids are a common reservoir, which means that the pathogen repeatedly pops up in dog populations. In other words: wherever people and unvaccinated, free-roaming dogs live close together, humans are at risk of rabies infection. Viral particles are shed abundantly in saliva, and as excessive salivation and aggression go hand-in-hand during the later stages of rabies, transmission usually occurs through bites from infected animals. However, the rabies virus can also enter the body through open wounds or through mucous membranes. In very rare cases, viral particles can be inhaled, but this typically only happens when particles are very densely concentrated in aerosols within closed environments, such as caves inhabited by thousands of infected bats.[1]

Globally, dog bites are the most common mode of transmission, with infected dogs responsible for about 99 percent of rabies deaths in people around the world. But in the United States, where dogs are protected by strict laws about rabies vaccines, canines account for less than 1 percent of reported rabies cases in people. Most rabies deaths in the US are instead caused by contact with infected bats, which are also rabies reservoirs, though in North America the disease is also frequently found in skunks and raccoons. But rabies infections from wildlife are more likely to be fatal in people when bats are the biters, in part because wounds from bats' needle-like teeth are so small that they often go unnoticed. The disease is then able to spread undetected until the first symptoms appear, and by then it's too late for treatment to do any good. (Needless to say, if you suspect that you or someone you know may have been exposed to rabies, put down this book *right now* and call a doctor immediately.)

While vaccines do prevent infection and can even stop rabies in its tracks if administered soon after exposure, rabies is almost always fatal once symptoms appear, which can take anywhere from a week to more than a year. Worldwide, rabies kills approximately 59,000 people annually—about half of which are children—mostly in rural parts of Africa, Asia, and Latin America. And before the disease paralyzes and kills its victims, it can reprogram their behavior.

After an animal—or a person—is infected, the virus first replicates at the site of transmission. It then travels along nerve cells through the body until it reaches the spinal cord, whereupon it goes straight to the brain. Once there, the virus multiplies further, collecting in the spaces between brain cells and inflaming brain tissue. It then migrates back into the body to the salivary glands and organs. Rabies infections in dogs and wildlife are associated with highly aggressive behavior, increased biting and snapping, and copious salivation (due to paralysis of the throat muscles), so diseased animals are sometimes described as "foaming at the mouth." Other symptoms that rabies causes in people include hallucinations, delirium, anxiety, fear of water—hydrophobia—and aerophobia, or fear of breezes and air drafts. Descriptions of people with these symptoms, usually sickened after encounters with infected dogs, appear in ancient texts from India and the Mediterranean Basin dating to thousands of years ago.[2]

Clinical signs of rabies and how long its symptoms last can be highly variable, and symptoms may appear sooner if the entry point for the virus is closer to the brain, such as a bite on the face. Soon after the virus invades the brain, the victim may enter the acute neural phase, known as the furious form of rabies. Heightened excitability, hyperactivity, and aggression are all associated with this stage in humans; such behaviors help to spread the virus. Eventually, the disease leads to cardiac and respiratory arrest.

But a victim may not exhibit the so-called furious symptoms at all during the final stage of illness. Instead, they may experience the paralytic form of rabies. This form appears in about 20 percent of human cases, according to the World Health Organization (WHO). It causes lethargy and paralysis, also culminating in death.[3]

Rabies manipulates behavior with the assistance of molecules called glycoproteins on the surface of the virus. For decades, these molecules were known to help the virus bind to nicotinic acetylcholine receptors found in muscles. And in 2017, scientists found that rabies' glycoproteins also enabled the virus to bind to the same receptors in brain cells and inhibit their activity—much in the same way that molecules in snake venom affect cell receptors. In experiments where a piece of the rabies glycoprotein was injected into the brains of mice, the mice became much more active than rodents that received a control injection; this type of hyperactivity is commonly seen in animals infected with rabies, the study authors reported. By derailing normal communication between brain cells, the virus could trigger behavior changes that favor transmission to new hosts.[4]

Though rabies is a dangerous zoonotic disease that poses a serious global threat to human health, we do have weapons on hand to prevent outbreaks from spiraling into anything resembling a zombie apocalypse. International agencies such as WHO are working to eliminate rabies, through vaccination programs and public education about recognizing the signs of rabies and avoiding contact with dogs and other animals that may be infected. One such WHO initiative in Bangladesh that administered vaccines to dogs in dozens of districts, led to a 50 percent decrease in human rabies deaths.[5] In the United States, the widespread availability of rabies vaccines for dogs helps keep pets safe if they encounter infected wildlife, and prevents the disease from spreading to people. After decades of effort by state and local health authorities to enforce dog vaccinations, in 2007 the Centers for Disease Control and Prevention (CDC) declared the US to be canine-rabies free (meaning that dogs were no longer reservoirs for the disease). This achievement was "one of the major public health success stories in the past 50 years," Charles Rupprecht, chief of the CDC Rabies Program, said at the time.

"Our public health infrastructure, including our quarantine stations, local animal control programs, veterinarians, and clinicians all play a vital role in preserving the canine-rabies-free status in the US," said Rupprecht. What's more, he added, installing and supporting such infrastructure in other nations could lead to similar success stories.[6]

But while dog-to-dog transmission of the disease had been all but eliminated in the US, Rupprecht warned, American dog owners still need to continue vaccinating their pets. Today, in the US, approximately 4,000 animals—mostly bats, foxes, raccoons, and skunks—still test positive for rabies each year, according to the USDA Animal and Plant Health Inspection Service.[7]

"Rabies is ever-present in wildlife and can be transmitted to dogs or other pets," Rupprecht said. "We need to stay vigilant."

Researchers are also working to develop treatments for rabies. In 2023, scientists at the Uniformed Services University (USU), a health science institution of the US federal government in Bethesda, Maryland, reported a promising breakthrough in the form of a monoclonal antibody, a lab-made molecule that mimics or boosts the body's immune response. The treatment, which was named F11, was tested on two strains of *Lyssavirus*: Australian bat lyssavirus (ABLV) and rabies virus (RABV). In experiments with infected mice, a single dose of F11 slammed the brakes on viral replication during the early stages of infection. To the researchers' surprise, the treatment had another, unexpected effect. Sick mice that received a dose of F11 survived their infections, even if the dose was given after the virus had already reached the host's central nervous system and was causing neurological symptoms—by then, death is almost always inevitable. The treatment was equally effective on both viral strains.[8]

What made this so astonishing is that monoclonal antibodies typically can't cross the blood-brain barrier, a membrane that prevents large molecules and antibodies from entering the brain and spinal cord (this is what hinders the effectiveness of traditional vaccines once rabies hits the central nervous system). When the scientists investigated further, they found that while F11 wasn't able to directly reach the brain, it was able to change certain immune cells in the brain—CD4-positive T cells—so that they were able to successfully attack the *Lyssavirus* infection themselves. If CD4-positive T cells were absent in an infected mouse subject, a dose of F11 was unable to save its life.

Should this treatment prove to be effective against rabies in humans, it could also be applied to neurological diseases caused by other types of viruses, said Brian Schaefer, a professor in USU's Department of Microbiology and Immunology and senior author of the study describing the F11 treatment, published in the journal *EMBO Molecular Medicine*.

"If these findings can be generalized to other viral pathogens that target the spinal cord and brain," said Schaefer, "this work may have broad implications for treating many other human viral diseases of the central nervous system."

*

Rabies isn't the only microbial pathogen associated with behavior manipulation in mammals. And if you're a cat owner, you'd better brace yourself: this mind-controlling microbe may already be a guest in your home.

Toxoplasma gondii is a parasitic protozoan, a single-celled organism that's related to the malaria parasite (*Plasmodium berghei*) and reproduces only in the intestines of felines, their definitive hosts. Recent estimates based on 50 years of data suggest that in some parts of the world, *T. gondii* can be found in at least 50 percent of wild felines and is present in about 30 percent or more of domestic cats.[9] Infected cats excrete noninfective, thick-walled time bombs of *T. gondii* infection called oocysts, and cats can shed millions of oocysts in feces per day for up to three weeks after acquiring the parasite. Oocysts become infectious within a few days after leaving the cat's body, and they can survive for a year or more in soil and water. They can then be picked up by intermediate hosts, and at least 350 species of birds and mammals—humans included—are potential carriers. Inside these secondary hosts, the parasite progresses to a fast-growing life stage known as a tachyzoite. It burrows into tissues of the brain, nerves, and muscles, where it becomes a bradyzoite, or tissue cyst. When it multiplies there, it causes the disease toxoplasmosis (its feline hosts are spared the disease—acute cases of toxoplasmosis in cats are extremely rare).

But in order for *T. gondii* to make more *T. gondii*, it needs to return to a cat's gut. How exactly does it do that from within a different animal species? One solution is to manipulate that host's behavior, so that its bradyzoite-stuffed body tissues will end up being eaten by a cat.

Rodents are frequent intermediate hosts for *T. gondii*; they acquire the parasite directly from infected cat feces or from contact with contaminated environments. A healthy mouse or rat is usually wary of any sign that a cat might be nearby; they shun the scent of cat urine and generally give felines a wide berth. But when rats and mice are infected with *T. gondii*, the rodents become unnaturally bold around cats. Within a few weeks of infection, the parasite takes the rodents' natural aversion to pheromones in cat urine and spins it into attraction. Studies have suggested that this may be linked to the presence of parasitic cysts in the amygdala, the brain region that regulates fear. There, *T. gondii* causes retraction in dendrites, the branching structures in neurons that communicate information to other nerve cells. Production of the hormone corticosterone, which is linked to stress responses, is also reduced. In healthy rats, the odor of cat urine is known to trigger specific brain circuits in one part of the rat amygdala. But researchers in Singapore and the United States discovered that in infected adult male rats, olfactory cues from cat urine hyperactivated neurons in a different part of the rat amygdala—the region that typically perks up in response to sociosexual cues, such as hormones linked to mating and reproduction.[10]

T. gondii infection can also lead to other hormonal changes such as increased production of dopamine and testosterone. This too can dial down a host's fear of the unfamiliar, dial up its aggression, and prompt hyperactivity and risky behavior, such as ignoring scent cues from predators.[11]

And an infected mouse's newfound boldness extends beyond cats. In 2020, researchers at the University of Geneva and the University of Toronto found that mice harboring *T. gondii* were also attracted to the scent of urine from foxes and guinea pigs. Infected mice were also less anxious under circumstances that would frighten uninfected mice. They explored exposed sections of a laboratory maze, which uninfected mice avoided. They showed no fear when introduced to a live anesthetized rat, a frequent predator of mice (when not anesthetized, that is). And when one of the experimenters placed their hand near the wall of the mice's enclosure, infected mice approached and even briefly head-bumped the invading hand. They "thoroughly interacted with it" while their healthy cagemates cowered at a distance, researchers wrote in the journal *Cell Reports*.[12]

"These findings refute the myth of a selective loss of cat fear in *T. gondii*-infected mice, and point toward widespread immune-related alterations of behaviors," the study authors said. Under *T. gondii*'s influence, the scientists reported, signals get scrambled in the rodent brain when it faces a potential threat.

Infection also affected normal attraction in mice, the researchers discovered. For example, a healthy mouse would sooner approach another mouse than interact with an inanimate object. Not so in the case of infected mice; they were still curious about other mice but were equally inquisitive when presented with a metal cube or an apple. Changes in the mice's behavior were more dramatic when there were more cysts present in their brain tissue and when infection caused more brain inflammation, the researchers found.

"Taken together, these findings suggest that chronic *T. gondii* infection reduces anxiety and risk aversion while increasing curiosity and exploratory behavior," said co–first author Madlaina

Boillat, a postdoctoral fellow in neurogenetics at the University of Geneva.[13]

This brain rewiring might even be a permanent condition; scientists learned in 2013 that behavioral changes persisted in mice for months after an acute infection cleared up. Mice that had been infected with *T. gondii* were still attracted to cat urine, even after the parasite was undetectable in their bodies and there was no lingering evidence of inflammation in their brains.[14]

Odd behavior in *T. gondii* intermediate hosts isn't limited to rodents, either. Gray wolves (*Canis lupus*) that live alongside cougars (*Puma concolor*) in Wyoming's Yellowstone National Park became more aggressive and more risk-averse pack leaders when infected with *T. gondii*. Infected captive chimpanzees (*Pan troglodytes troglodytes*) in Gabon were attracted to the urine of leopards (*Panthera pardus*), which are the only natural predators of chimps. In the Maasai Mara National Reserve in Kenya, infected spotted hyena cubs (*Crocuta crocuta*) were bolder around lions (*Panthera leo*), approaching much closer to the big cats than uninfected cubs did—and with that boldness came a much higher mortality rate, according to researchers from the United States and Kenya who reviewed three decades of data on hyenas living in the reserve.

"Among hyenas infected as cubs, 100% of the deaths were caused by lions," the scientists reported in 2021 in the journal *Nature Communications*, "while only 17% of the deaths of hyenas not infected as cubs were caused by lions."[15]

Behavior changes in nonrodent hosts are not restricted to reduced fear of felines. In California sea otters, also known as southern sea otters (*Enhydra lutris nereis*), *T. gondii* infection can cause a disease of the central nervous system called toxoplasmic encephalitis. Otters with even moderate cases of the illness are nearly four times more likely than healthy otters to be attacked by sharks, perhaps because they are bolder around these predators than uninfected otters, researchers suggested.[16]

And then there's the impact of *T. gondii* on people.

Humans are what is known as "dead-end hosts" for *T. gondii.* The parasite can infect people, but that doesn't do it any good when it comes to reproduction; once *T. gondii* gets inside a person, chances are slim that it will ever return to a feline primary host. (The odds may have been more in the parasite's favor in our evolutionary past; bite marks on fossils suggest that early humans and our Neanderthal cousins were prey or were scavenged by saber-toothed cats. Though, in some parts of the world today, humans that live near the natural habitats of big cats such as tigers, lions, and cougars still run the risk of fatal attacks.) Nevertheless, there's a growing body of evidence that modern humans are vulnerable to *T. gondii* manipulation that manifests in a variety of behaviors—though, none of those behaviors is likely to result in the host being eaten by a cat. (Probably.)

T. gondii is surprisingly common in people, with more than 2 billion people worldwide estimated to be chronically infected. In the United States, more than 40 million people may be carriers of *T. gondii,* and toxoplasmosis is one of the leading causes of death in cases of foodborne illnesses, according to the CDC.[17] Transmission can be zoonotic, through direct contact with infected cats and their feces (oocysts become infectious within one to five days after being excreted). It may be acquired by eating contaminated meat, fish, or poultry that is undercooked, or by drinking unpasteurized milk. It can also be congenital, transmitted to a fetus during pregnancy. In very rare cases, a person may be infected via an organ transplant or blood transfusion from an infected donor.

The good news is that in the vast majority of human cases, *T. gondii* doesn't cause illness. Infection is usually kept under control by a healthy immune system, and though the parasite isn't eradicated in the host's body, a person who carries *T. gondii* may never know that they're infected. But in people with compromised immune systems and in pregnant people, the parasite is more likely to evade an immune response and multiply, and the symptoms of toxoplasmosis can be severe. Mild infections can cause swollen lymph glands and flu-like pain such as muscle aches, which can last for more than a month. In severe cases, toxoplasmosis damages the brain, nervous system, eyes, and other organs, and in pregnant people it can cause miscarriages. Infants that were exposed to *T. gondii* during pregnancy may show no signs of illness at birth but are at risk for developing toxoplasmosis later in life, potentially leading to blindness and seizures.

Toxoplasmosis can be detected through blood tests, and there are treatments to keep symptoms at bay. However, in patients with weakened immune systems, it may not be possible to eliminate the parasite entirely, and medication may be required for as long as they are immunosuppressed.

While public health programs typically focus on addressing active infections of *T. gondii*, latent infection may also shape human behavior. And if a picture of a "crazy cat lady" just popped into your head, you wouldn't be the first to make that connection. News coverage of *T. gondii* often links the parasite to the time-honored and persistent stereotype of an eccentric single woman surrounded by felines. However, this trope likely has more to do with systemic misogyny and historically derogatory portraits of women who live alone and prefer cats to the company of men than it does with the brain-controlling powers of a microbial parasite. Despite what headlines may imply, owning cats isn't a one-way ticket to psychosis.

That said, evidence suggests that a latent *T. gondii* infection can affect the human mind. For decades, scientists have been investigating the association between *T. gondii* and mental health, though experts are still piecing together exactly how the parasite disrupts the human central nervous system and contributes to mental disorders, alongside other factors such as an individual's age, genetics, and overall health. One possibility is that *T. gondii* triggers immune responses that hamper communication between neurons, tweaking the production of neurotransmitters that regulate stress responses, such as dopamine, serotonin, and norepinephrine. Another is by causing brain inflammation through the presence of cysts in brain tissue.

Dozens of studies published over the past 20 years suggest that *T. gondii* infection poses a risk factor for the development of schizophrenia, independent of genetic factors. *T. gondii* has also been associated with heightened risk for generalized anxiety disorder, as well as obsessive-compulsive disorder, bipolar disorder, depression, and suicidal behavior.[18] An analysis of 13 studies that was recently conducted by Dutch researchers in the University of Amsterdam's Department of Psychiatry, found that otherwise healthy patients with latent *T. gondii* infections didn't perform as well in cognitive tests as uninfected subjects did.[19] Latent toxoplasmosis has even been associated with a greater risk of being in a traffic accident, perhaps because the infection affects the host's reaction time and concentration, according to researchers in the Czech Republic.[20]

By manipulating hormone output, *T. gondii* may also be responsible for causing personality changes. For instance, testosterone is known to affect impulsiveness, aggression, and risk-taking, though people may respond differently to infection, depending on factors, such as gender, that brew the hormonal soup in our bodies. In France—where up to 43 percent of the population is estimated to carry the *T. gondii* parasite—scientists at the Université de Montpellier have suggested that "infected men appear to be more dogmatic, less confident, more jealous, more cautious, less impulsive and more orderly than others." Infected women, by comparison, "seem warmest, more conscientious, more insecure, more sanctimonious and more persistent than others," the researchers reported.[21]

Certain political values may be amplified by *T. gondii*, too. In 2022, researchers with the Czech Academy of Sciences and Charles University in Prague suggested that people with latent infections leaned toward tribalism and away from anti-authoritarianism.[22] An increased tendency toward risky behavior may also affect economic decision-making in infected people. When scientists from the United States, Norway, Spain, and Hong Kong investigated how university students might be affected by latent *T. gondii* infections, they found that students who tested positive for *T. gondii* were 1.4 times more likely to major in business. Infected students were also 1.7 times more likely than uninfected ones to focus on high-risk ventures, such as entrepreneurship, rather than on subjects that were more stable and predictable: accounting or marketing, for example. After the researchers visited entrepreneurship events and tested adult professionals, they found that infected attendees were 1.8 times more likely than their colleagues to have launched their own businesses. The scientists then reviewed more than two decades of business data from 42 countries, alongside data about prevalence of *T. gondii* infection in the general public. They found a correlation: countries with higher instances of infection also had higher rates of entrepreneurship.

Because new business ventures often fail, it's reasonable to expect entrepreneurship to be accompanied by a fear of failure. But when researchers looked at *T. gondii* numbers, they saw that the opposite was true; in countries where *T. gondii* infection was higher, fewer entrepreneurs self-reported that they feared failing. "A fear of failure is quite rational," said lead study author Stefanie K. Johnson, an associate professor in Colorado University (CU) Boulder's Leeds School of Business. "*T. gondii* might just reduce that rational fear."[23]

As long as humans have been around, our civilizations, culture, and history have been shaped by the influence of infectious diseases. As scientists unravel the mechanisms of microbes like *T. gondii*, the closer they get to understanding how deep that manipulation goes; and the impact of these manipulators on people may be even greater than we suspect, said Pieter Johnson, study coauthor and a professor in CU Boulder's Department of Ecology and Evolutionary Biology.

"As humans, we like to think that we are in control of our actions," Johnson said. "But emerging research shows that the microorganisms we encounter in our daily lives have the potential to influence their hosts in significant ways."[24]

*

In reality, neither rabies nor toxoplasmosis is likely to produce armies of human zombies like the mind-controlled ghouls that lurch across our movie screens, televisions, and computer monitors. But that type of zombie outbreak has nonetheless been a source of inspiration for scientists, medical officials, and government health agencies who use zombies to promote messages about disaster response. It certainly inspired the CDC in May 2011, when they debuted "Preparedness 101: Zombie Apocalypse" on the agency's blog Public Health Matters.

"The rise of zombies in pop culture has given credence to the idea that a zombie apocalypse could happen," wrote physician Ali S. Khan, an assistant surgeon general, who at the time led the CDC's Office of Public Health Preparedness and Response. "The proliferation of this idea has led many people to wonder 'How do I prepare for a zombie apocalypse?' Well, we're here to answer that question for you, and hopefully share a few tips about preparing for *real* emergencies too!"[25]

Communications staff at the CDC came up with the idea after seeing engagement on Twitter (now X) spike around a thread that mentioned zombies, according to Dave Daigle, a CDC spokesperson and leader of the campaign. The CDC declared that, as with any type of natural disaster (like a flood or earthquake), safely weathering zombie attacks requires preparation, such as assembling an emergency kit and planning an evacuation route. Khan outlined what a CDC response might be, should zombies start roaming the streets (see figure 13).

Figure 13 A zombie-themed poster produced by the CDC hints at the importance of being prepared for disasters and emergencies. *Courtesy of the Centers for Disease Control and Prevention*

"CDC would conduct an investigation much like any other disease outbreak," Khan said. An investigation would have several goals: "determine the cause of the illness, the source of the infection/virus/toxin, learn how it is transmitted and how readily it is spread, how to break the cycle of transmission and thus prevent further cases, and how patients can best be treated." The CDC even produced a downloadable graphic novel on the topic, complete with a checklist that would prepare readers "for any kind of disaster, even zombies."

The campaign, which at the time of launch was so popular that it crashed the CDC's website, has since been retired. But its success spawned many imitations. The American Red Cross of Massachusetts crafted a similar campaign, which later went national. The Canadian Red Cross also chimed in, suggesting on its website that tools in an emergency kit could be useful "in the event of an outbreak of the zombie virus."[26] Medical training sessions at hospitals have tested zombie-themed escape rooms, in which nurses and trainees are tasked with treating victims and containing infection on a tight timeline, against a backdrop of a zombie outbreak. Even the Federal Emergency Management Agency (FEMA) added zombies into messages about disaster preparedness, partnering in 2019 with the producers and cast of the film *Zombieland: Double Tap*. Posters and videos for the campaign shared the slogan: "Zombies don't plan ahead. You can. Make your emergency plan."[27]

Zombies and zombification—in research and in science communication—also fuel the work of Arizona State University's so-called zombie professor. Athena Aktipis is an associate professor in ASU's Department of Psychology, and is director of The Cooperation Lab, which investigates how human cooperation is shaped by culture and biology, and how microbes influence human behavior. Aktipis also hosts the podcast *Zombified*, about all the ways that our brains can be hijacked by external factors, and is the cofounder and chair of a biannual conference known as the Zombie Apocalypse Medicine Meeting, or ZAMM.

Launched at ASU in 2018, ZAMM was designed to bring together scientists from a variety of disciplines. Aktipis and the other organizers decided to present and discuss biological and social science research through the lens of a zombie apocalypse "because that'll make it much more fun to talk about parasites, and infectious diseases, and all the ways that we zombify each other," Aktipis says.

"It started out as a joke," she recalls. "And the next day we woke up and were like, 'Actually, maybe we should do that.' "

With zombies as a central theme, sessions at the meeting touch on diverse issues, such as microbial mind-manipulation and zombification at the cellular level, as well as community resilience and prioritizing healthcare access during outbreaks, to name a few. Other panels addressed how people are "zombified" by personal devices and autonomous technology; by personal beliefs and relationships; and by our own immune systems.

"Zombies work on a lot of different levels for dealing with complex topics," says Aktipis. "Once you're talking about zombies, you can switch between the more metaphorical and the more literal, making it easier to have an interdisciplinary approach."

But by the time the next ZAMM meeting rolled around in 2020, an actual global threat had emerged that temporarily brought the social machinery of everyday life to a grinding halt, overwhelming healthcare responses as thoroughly as any fictional zombie plague would. The COVID-19 pandemic, caused by the highly infectious SARS-CoV-2 coronavirus, first appeared toward the end of 2019 and spread like wildfire around the world. The virus causes acute respiratory distress and has been linked to a range of neurological and cardiovascular symptoms that can persist for years after infection. As of January 2024, the World Health Organization (WHO) estimates that the virus has infected nearly 800 million people globally, and caused at least 7 million deaths (the actual figure is likely much higher, WHO says, as mortality caused by COVID-19 has been significantly underreported in many countries).[28] Vaccines are now widely available, but organized public safety measures to contain the disease have largely been abandoned. With COVID-19 still freely circulating in most of the world, new strains continue to evolve, sickening millions of people.

For Aktipis, the speedy transmission of COVID-19 recalled the swift and deadly spread of a zombifying infection. In fact, SARS-CoV-2 further resembles a fictional zombie virus (and other behavior-changing pathogens) because it affects how people behave when they're infected, in order to increase the rates of transmission, Aktipis wrote for *The Conversation* in 2021.[29] Upon infection, the coronavirus disrupts the normal immune response and blocks interferons, the molecules that are your body's first line of defense against viruses. The virus also suppresses activity in cytokines, another type of molecule that regulates an immune response to repel invaders. When these molecular signals are activated, they trigger behavior changes. Sick people become more withdrawn, are less active, and less social. But in the early stages of COVID-19 infection, infected people typically continue with their normal routines, spreading the virus for days before they start feeling sick.

And about 40 percent of people infected by SARS-CoV-2 never develop symptoms; they too become unknowing agents for spreading the virus far and wide.

"It certainly made me think about zombies," Aktipis says. "Maybe the zombie apocalypse is already upon us."

*

"This isn't going to end well."

These six words are matter-of-factly uttered in the film *The Dead Don't Die* by Officer Ronnie Peterson (Adam Driver) when he realizes that a zombie outbreak threatens his town. Peterson's perspective isn't unusual; uninfected people in most zombie stories are usually pessimistic about the outcome. But if you've read this far, I hope that by now you've gained some perspective on zombification—and you might even be starting to see its benefits. Zombifying organisms can keep certain insect populations in check, so they don't outcompete other arthropods. Their host-specific puppeteering means that they can be used to control invasive species and curb the spread of agricultural pests without poisoning every arthropod in sight. Compounds that zombifiers use to aid them in their mind-controlling habits are being studied for medical applications; venom in parasitic wasps that zombify roaches, for example, may hold clues for treating diseases of the central nervous system, such as Parkinson's. And molecules produced by the zombie ant fungus *Ophiocordyceps unilateralis* and related species, which interact with a host's immune system and stimulate its metabolism to produce antibacterial agents, may prove useful for developing medicines that reduce inflammation, curb infections, and target cancer cells.

Even for scientists who don't directly research zombifying parasites and their victims, zombification can still serve as a source of inspiration. Creative science communicators at ZAMM, the CDC, the Red Cross, and various hospitals have all used zombies as proxies. Zombie scenarios can even be a source of important scientific data.

For example, researchers at Argonne National Laboratory in Illinois calculated that in the event of a zombie outbreak, Chicago would be completely overwhelmed by the undead in under two months, according to a computational model using data from highly infectious diseases such as Ebola and methicillin-resistant *Staphylococcus aureus* (MRSA). This training exercise for their model, the scientists said, prepares the model for addressing a real health crisis, in which it would guide health officials as they plan and coordinate a response.[30]

And at Aalto University in Finland, when another team of researchers recently modeled a zombie outbreak across the nation, they were astonished to find how swiftly contagion spread. A scenario starting with just one zombie in Helsinki showed that officials would have a mere seven hours to contain the spread, either by eliminating the zombies or quarantining the capital. If containment failed, the entire country would be overrun by the escaping undead.

"I shouldn't have found it surprising, but I was surprised at how quickly we have to react to keep our population alive," said Pauliina Ilmonen, leader of the Finnish study and an associate professor in Aalto University's Department of Mathematics and Systems Analysis. "It made me think about moral issues like the rights of individuals versus the rights of a population."[31]

Meanwhile, for insects and arthropods around the world, the ongoing zombie apocalypse continues. Somewhere in a forest right now, an ant is staggering up a leaf and snapping its jaws tight around a vein. In a suburban backyard, a cricket is hopping into a swimming pool and paddling frantically to keep afloat, as a threadlike worm unspools from its backside. Atop a garden bed, a caterpillar is rearing up to defend a brood of wasp larvae in cocoons that sprout from its back. These zombies, and the uncounted millions that came before them, never knew what it was that stole their will and their lives—but we do. And the more that we learn about these zombifiers, the more we will discover about evolution, the codependence of parasites and hosts, and even about our own health and behavior.

As the CDC once wisely warned: "Don't be a zombie; be prepared." In the natural world, in epidemiology, in pop culture, and in our imaginations—zombies aren't going anywhere. We may as well embrace them.

Or at least learn from them.

Acknowledgments

As this zombie tale lurches to an end, I'd be making a grave mistake if I didn't mention all the marvelous brains that helped keep this project alive and kicking. At the top of the list is senior acquisitions editor Tiffany Gasbarrini at Hopkins Press—this book wouldn't exist without her enthusiasm and guidance. Thanks as well to assistant acquisitions editor Ezra Rodriguez, production editor Hilary Jacqmin, copyeditor Susan Matheson, and the rest of the talented crew at the Press who shepherded a gooey blob of a first draft to the exquisite final form you hold in your hands.

Tons of gratitude to Henry Bloomstein and Jacqueline Green, who offered thoughtful words of wisdom for navigating the sticky world of book writing when I didn't even know which questions to ask, let alone how to answer them. Thanks as well to Greg Gbur, Bill Schutt, and Harold Goldberg for their bookish wisdom; to Kristin Hugo and her monthly Authors of Nonfiction Books in Progress meeting, which provided a much-needed lifeline to other writers (and *loads* of helpful advice); and to Geralyn Abinader and Paula Schaap for their critique and feedback.

A heartfelt thank you to all of the scientists who generously took the time to field my questions; I hope that I've done justice to their explanations and expertise. Any errors here are mine, not theirs.

I'm beyond grateful to my parents, Joshua and Rachelle Weisberger, whose love and support has been constant and unflagging.

Finally, to Hens and Martin—I couldn't have done this without you and I wouldn't have wanted to. You make the good things great, the not-so-good things bearable, and everything better. You are my favorite nerds and my love for you is undying.

Notes

Introduction

1. "Fossil Reveals 48 Million Year History of Zombie-Ants," University of Exeter, News Archive, August 18, 2010, https://www.exeter.ac.uk/news/archive/2010/august/title_96180_en.html.
2. Connor J. Meyer, Kira A. Cassidy, Erin E. Stahler, Ellen E. Brandell, Colby B. Anton, Daniel R. Stahler, and Douglas W. Smith, "Parasitic Infection Increases Risk-Taking in a Social, Intermediate Host Carnivore," *Communications Biology* 5 (November 24, 2022): 1180, https://doi.org/10.1038/s42003-022-04122-0.
3. Eben Gering, Zachary M. Laubach, P.S.D. Weber, Gisela Soboll Hussey, Kenna D. S. Lehmann, Tracy M. Montgomery, Julie W. Turner et al., "*Toxoplasma Gondii* Infections Are Associated with Costly Boldness toward Felids in a Wild Host," *Nature Communications* 12 (June 22, 2021): 3842, https://doi.org/10.1038/s41467-021-24092-x.
4. John Easton, "What Does It Mean When 2 Billion People Share Their Brain with a Parasite?," UChicago Medicine, July 22, 2014, https://www.uchicagomedicine.org/forefront/biological-sciences-articles/what-does-it-mean-when-2-billion-people-share-their-brain-with-a-parasite.
5. US Centers for Disease Control and Prevention, "Toxoplasmosis: About Toxoplasmosis," https://www.cdc.gov/toxoplasmosis/about/.

Chapter 1. Zombifiers

1. Zhifei Zhang, Luke C. Strotz, Timothy P. Topper, Feiyang Chen, Yanlong Chen, Yue Liang, Zhiliang Zhang et al., "An Encrusting Kleptoparasite-host Interaction from the Early Cambrian," *Nature Communications* 11 (June 2, 2020): 2625, https://doi.org/10.1038/s41467-020-16332-3.
2. Mindy Weisberger, "Cambrian Fossils Show Oldest Example of Parasites in Action," Livescience.com, June 2, 2020, https://www.livescience.com/cambrian-parasites.html.
3. Daniel Lima, Damares Ribeiro Alencar, William Santana, Naiara C. Oliveira, Antônio Álamo Feitosa Saraiva, Gustavo R. Oliveira, Christopher B. Boyko et al., "110-Million-Years-Old Fossil Suggests Early

Parasitism in Shrimps," *Scientific Reports* 13 (September 4, 2023): 14549, https://doi.org/10.1038/s41598-023-40554-2.

4. George O. Poinar and Yves-Marie Maltier, "*Allocordyceps baltica* gen. et sp. nov. (Hypocreales: Clavicipitaceae), an Ancient Fungal Parasite of an Ant in Baltic Amber," *Fungal Biology* 125, no. 11 (November 2021): 886–90, https://doi.org/10.1016/j.funbio.2021.06.002.
5. Jill Langlois, "First Evidence of Parasites in Dinosaur Bones Found," Smithsonian Magazine, November 18, 2020, https://www.smithsonianmag.com/science-nature/first-evidence-parasites-dinosaur-bones-found-180976309/.
6. David L. Reed, Jessica E. Light, Julie M. Allen, and Jeremy J. Kirchman, "Pair of Lice Lost or Parasites Regained: The Evolutionary History of Anthropoid Primate Lice," *BMC Biology* 5 (March 7, 2007): 7, https://doi.org/10.1186/1741-7007-5-7.
7. Kelly L. Weinersmith, Sean M. Liu, Andrew A. Forbes, and Scott P. Egan, "Tales from the Crypt: A Parasitoid Manipulates the Behaviour of Its Parasite Host," *Proceedings of the Royal Society B: Biological Sciences* 284, no. 1847 (January 25, 2017): 20162365, https://doi.org/10.1098/rspb.2016.2365.
8. Sara B. Weinstein and Armand M. Kuris, "Independent Origins of Parasitism in Animalia," *Biology Letters* 12, no. 7 (July 2016): 20160324, https://doi.org/10.1098/rsbl.2016.0324.
9. Andy Dobson, Kevin D. Lafferty, Armand M. Kuris, Ryan F. Hechinger, and Walter Jetz, "Homage to Linnaeus: How Many Parasites? How Many Hosts?," in *Biodiversity and Extinction*, vol. 2 of *The Light of Evolution*, ed. J. C. Avise, S. P. Hubbell, and F. J. Ayala (Washington, DC: National Academies Press, 2008), https://www.ncbi.nlm.nih.gov/books/NBK214895/.
10. "Parasitoid Wasps: Indispensable Insects You Never Think About . . . Or Never Want To!," Darwin Tree of Life, February 15, 2022, https://www.darwintreeoflife.org/news_item/parasitoid-wasps-indispensable-insects-you-never-think-about-or-never-want-to/.
11. "Strepsiptera: Twisted Wing Flies," Royal Entomological Society, accessed June 8, 2024, https://www.royensoc.co.uk/understanding-insects/classification-of-insects/strepsiptera/.
12. Jay Sullivan, "Unwanted Guests: The Weird World of Parasitic Plants," Natural History Museum, accessed June 8, 2024, https://www.nhm.ac.uk/discover/parasitic-plants.html.
13. Shelley A. Adamo, "Turning Your Victim into a Collaborator: Exploitation of Insect Behavioral Control Systems by Parasitic Manipulators," *Current Opinion in Insect Science* 33 (June 2019): 25–29, https://doi.org/10.1016/j.cois.2019.01.004.
14. Eloise B. Cram, "Developmental Stages of Some Nematodes of the Spiruroidea Parasitic in Poultry and Game Birds," *Technical Bulletin of US Department of Agriculture* 227 (February 1931): 1–27.

15. J. C. Holmes and W. M. Bethel, "Modification of Intermediate Host Behaviour by Parasites," in *Behavioural Aspects of Parasite Transmission*, ed. Elizabeth U. Canning and C. A. Wright (New York: Academic Press, 1972): 123–49, https://www.cabidigitallibrary.org/doi/full/10.5555/19730804260.
16. Janice Moore, "Parasites That Change the Behavior of Their Host," *Scientific American* 250, no. 5 (May 1984): 108, https://www.scientificamerican.com/article/parasites-that-change-the-behavior/.
17. D. P. Hughes, S. B. Andersen, N. L. Hywel-Jones, W. Himaman, J. Billen, and J. J. Boomsma, "Behavioral Mechanisms and Morphological Symptoms of Zombie Ants Dying from Fungal Infection," *BMC Ecology* 11 (May 9, 2011): 13, https://doi.org/10.1186/1472-6785-11-13.
18. Janice Moore, "The Behavior of Parasitized Animals," *BioScience* 45, no. 2 (February 1995): 89–96, https://doi.org/10.2307/1312610.

Chapter 2. Fungus Among Us, Part I

1. Hui-Chen Lo, Chienyan Hsieh, Frank Yeong-Sung Lin, and Tai-Hao Hsu, "A Systematic Review of the Mysterious Caterpillar Fungus *Ophiocordyceps sinensis* in DongChongXiaCao (冬蟲夏草 Dōng Chóng Xià Cǎo) and Related Bioactive Ingredients," *Journal of Traditional and Complementary Medicine* 3, no. 1 (January 2013): 16–32, https://doi.org/10.1016/s2225-4110(16)30164-x.
2. "Cordyceps: Attack of the Killer Fungi—Planet Earth Attenborough BBC Wildlife," BBC Studios, video, 3:03, November 3, 2008, https://www.youtube.com/watch?v=XuKjBIBBAL8.
3. Beth Accomando, "A Zombie Horror Game, Inspired by . . . a Nature Documentary?," NPR, July 9, 2013, https://www.npr.org/2013/07/09/199040676/a-zombie-horror-game-inspired-by-a-nature-documentary.
4. Maridel Fredericksen, Yizhe Zhang, Missy L. Hazen, Raquel G. Loreto, Colleen A. Mangold, Danny Z. Chen, and David Hughes, "Three-Dimensional Visualization and a Deep-Learning Model Reveal Complex Fungal Parasite Networks in Behaviorally Manipulated Ants," *Proceedings of the National Academy of Sciences* 114, no. 47 (November 7, 2017): 12590–95, https://doi.org/10.1073/pnas.1711673114.
5. George Beccaloni, "Sulawesi: Wallace's Laboratory of Evolution," The Alfred Russel Wallace Website, August 2019, https://wallacefund.myspecies.info/content/sulawesi-wallaces-laboratory-evolution.
6. Susan Goldhor, "Alfred Russel Wallace: The Fungal Connection," *FUNGI* 7, no. 5 (Winter 2014): 34–36, https://fungimag.com/archives/v7n5_winter_2014.htm.
7. David Hughes, Sandra Breum Andersen, Nigel L. Hywel-Jones, Winanda Himaman, Johan Billen, and Jacobus J. Boomsma, "Behavioral Mechanisms and Morphological Symptoms of Zombie Ants Dying from Fungal Infection," *BMC Ecology* 11 (2011): 13, https://doi.org/10.1186/1472-6785-11-13.

8. João P. M. Araújo, H. C. Evans, Ryan M. Kepler, and David Hughes, "Zombie-Ant Fungi across Continents: 15 New Species and New Combinations within *Ophiocordyceps*. I. Myrmecophilous Hirsutelloid Species," *Studies in Mycology* 90, no. 1 (June 2018): 119–60, https://doi .org/10.1016/j.simyco.2017.12.002.
9. "Discovery about Evolution of Fungi Has Implications for Humans, Says U of M Researcher," EurekAlert!, American Association for the Advancement of Science, October 20, 2006, https://www.eurekalert.org /news-releases/764806.
10. Kevin D. Hyde, "The Numbers of Fungi," *Fungal Diversity* 114, no. 1 (May 2022): 1, https://doi.org/10.1007/s13225-022-00507-y.
11. Araújo et al., "Zombie-Ant Fungi across Continents."
12. Gi Ho Sung, Nigel L. Hywel-Jones, Jae Mo Sung, J. Jennifer Luangsaard, Bhushan Shrestha, and Joseph W. Spatafora, "Phylogenetic Classification of *Cordyceps* and the Clavicipitaceous Fungi," *Studies in Mycology* 57, no. 1 (January 2007): 5–59, https://doi.org/10.3114/sim .2007.57.01.
13. Harry C. Evans, Simon L. Elliot, and David P. Hughes, "*Ophiocordyceps unilateralis*: A Keystone Species for Unraveling Ecosystem Functioning and Biodiversity of Fungi in Tropical Forests?," *Communicative & Integrative Biology* 5, no. 4 (September 1, 2011): 598–602, https://doi .org/10.4161/cib.16721.
14. Araújo et al., "Zombie-Ant Fungi across Continents."
15. Wei Lin, Yung I. Lee, Shao Lun Liu, Chih-Chieh Lin, Tan Ya Chung, and Jui Yu Chou, "Evaluating the Tradeoffs of a Generalist Parasitoid Fungus, *Ophiocordyceps unilateralis*, on Different Sympatric Ant Hosts," *Scientific Reports* 10, no. 1 (April 14, 2020), https://doi.org/10 .1038/s41598-020-63400-1.
16. Charissa de Bekker, Lauren E. Quevillon, Philip B. Smith, Kimberly R. Fleming, Debashis Ghosh, Andrew D. Patterson, and David P. Hughes, "Species-Specific Ant Brain Manipulation by a Specialized Fungal Parasite," *BMC Evolutionary Biology* 14, no. 1 (August 29, 2014): 166, https://doi.org/10.1186/s12862-014-0166-3.
17. Charissa de Bekker, Robin A. Ohm, Raquel G. Loreto, Aswathy Sebastian, István Albert, Martha Merrow, Andreas Brachmann et al., "Gene Expression during Zombie Ant Biting Behavior Reflects the Complexity underlying Fungal Parasitic Behavioral Manipulation," *BMC Genomics* 16, no. 1 (August 19, 2015): 620, https://doi.org/10.1186/s12864-015 -1812-x.
18. Raquel G. Loreto and David Hughes, "The Metabolic Alteration and Apparent Preservation of the Zombie Ant Brain," *Journal of Insect Physiology* 118 (October 1, 2019): 103918, https://doi.org/10.1016/j .jinsphys.2019.103918.
19. William C. Beckerson, Courtney Krider, Umar A. Mohammad, and Charissa de Bekker, "28 Minutes Later: Investigating the Role of

Aflatrem-like Compounds in *Ophiocordyceps* Parasite Manipulation of Zombie Ants," *Animal Behaviour* 203 (September 2023): 225–40, https://doi.org/10.1016/j.anbehav.2023.06.011.

20. Gi-Ho Sung, George Poinar, and Joseph W. Spatafora, "The Oldest Fossil Evidence of Animal Parasitism by Fungi Supports a Cretaceous Diversification of Fungal–Arthropod Symbioses," *Molecular Phylogenetics and Evolution* 49, no. 2 (November 2008): 495–502, https://doi.org/10.1016/j.ympev.2008.08.028.
21. George O. Poinar and Yves-Marie Maltier, "*Allocordyceps Baltica* gen. et sp. nov. (Hypocreales: Clavicipitaceae), an Ancient Fungal Parasite of an Ant in Baltic Amber," *Fungal Biology* 125, no. 11 (November 2021): 886–90, https://doi.org/10.1016/j.funbio.2021.06.002.
22. David Hughes, Torsten Wappler, and Conrad C. Labandeira, "Ancient Death-Grip Leaf Scars Reveal Ant–Fungal Parasitism," *Biology Letters* 7, no. 1 (August 18, 2010): 67–70, https://doi.org/10.1098/rsbl.2010.0521.
23. João P. M. Araújo, Mitsuru Moriguchi, Sadao Uchiyama, Noriko Kinjo, and Yu Matsuura, "*Ophiocordyceps salganeicola*, a Parasite of Social Cockroaches in Japan and Insights into the Evolution of Other Closely-Related *Blattodea*-Associated Lineages," *IMA Fungus* 12 (February 5, 2021): 3, https://doi.org/10.1186/s43008-020-00053-9.
24. Patrick Schultheiss, Sabine S. Nooten, Runxi Wang, Mark K. L. Wong, François Brassard, and Benoit Guénard, "The Abundance, Biomass, and Distribution of Ants on Earth," *Proceedings of the National Academy of Sciences* 119, no. 40 (September 19, 2022): e2201550119, https://doi.org/10.1073/pnas.2201550119.
25. Hughes et al., "Behavioral Mechanisms and Morphological Symptoms."
26. Maj-Britt Pontoppidan, Winanda Himaman, Nigel L. Hywel-Jones, Jacobus J. Boomsma, and David P. Hughes, "Graveyards on the Move: The Spatio-Temporal Distribution of Dead *Ophiocordyceps*-Infected Ants," *PLOS ONE* 4, no. 3 (March 12, 2009): e4835, https://doi.org/10.1371/journal.pone.0004835.
27. Fernando Sarti Andriolli, Noemia Kazue Ishikawa, Ruby Vargas-Isla, Tiara Sousa Cabral, Charissa de Bekker, and Fabrício Beggiato Baccaro, "Do Zombie Ant Fungi Turn Their Hosts into Light Seekers?," *Behavioral Ecology* 30, no. 3 (May/June 2019): 609–16, https://doi.org/10.1093/beheco/ary198.
28. Raquel G. Loreto, Simon L. Elliot, Mayara L. R. Freitas, Thairine M. Pereira, and David P. Hughes, "Long-Term Disease Dynamics for a Specialized Parasite of Ant Societies: A Field Study," *PLOS ONE* 9, no. 8 (August 18, 2014): e103516, https://doi.org/10.1371/journal.pone.0103516.
29. João P. M. Araújo and David Hughes, "Zombie-Ant Fungi Emerged from Non-manipulating, Beetle-Infecting Ancestors," *Current Biology* 29, no. 21 (November 2019): 3735–3738.e2, https://doi.org/10.1016/j.cub.2019.09.004.

30. Tina Hesman Saey, "Just Warm Enough," Science News, August 8, 2019, https://www.sciencenews.org/article/just-warm-enough.

Chapter 3. Fungus Among Us, Part II

1. Carolyn Elya and Henrik H. de Fine Licht, "The Genus *Entomophthora*: Bringing the Insect Destroyers into the Twenty-First Century," *IMA Fungus* 12 (November 12, 2021): 34, https://doi.org/10.1186/s43008-021-00084-w.
2. Ferdinand Cohn, "*Empusa Muscae*: Die Krankheit der Stubenfliegen," Novorum Actorum, Academiae Caesareae Leopoldino-Carolinae Naturae Curiosorum, 1855, https://www.biodiversitylibrary.org/item/177216#page/435/mode/1up.
3. D. M. MacLeod, E. Müller-Kögler, and N. Wilding, "*Entomophthora* Species with *E. Muscae*-Like Conidia," *Mycologia* 68, no. 1 (1976): 1–29, https://doi.org/10.1080/00275514.1976.12019881.
4. Elya and de Fine Licht, "Genus *Entomophthora*."
5. Stuart B. Krasnoff, David Watson, Donna M. Gibson, and E. C. Kwan, "Behavioral Effects of the Entomopathogenic Fungus, *Entomophthora muscae* on Its Host *Musca domestica*: Postural Changes in Dying Hosts and Gated Pattern of Mortality," *Journal of Insect Physiology* 41, no. 10 (October 1995): 895–903, https://doi.org/10.1016/0022-1910(95)00026-q.
6. D. W. Watson, "*Entomophthora muscae*," Biological Control: A Guide to Natural Enemies in North America (website), Cornell University, College of Agriculture and Life Sciences, accessed June 10, 2024, https://biocontrol.entomology.cornell.edu/pathogens/entomophaga muscae.php.
7. Andreas Naundrup, Björn Bohman, Charles A. Kwadha, Annette Bruun Jensen, Paul G. Becher, and Henrik H. de Fine Licht, "Pathogenic Fungus Uses Volatiles to Entice Male Flies into Fatal Matings with Infected Female Cadavers," *ISME Journal* 16, no. 10 (October 2022): 2388–97, https://doi.org/10.1038/s41396-022-01284-x.
8. Patricia J. Brobyn and N. Wilding, "Invasive and Developmental Processes of *Entomophthora muscae* Infecting Houseflies (*Musca Domestica*)," *Transactions of the British Mycological Society* 80, no. 1 (February 1983): 1–8, https://doi.org/10.1016/s0007-1536(83)80157-0.
9. Carolyn Elya, Tin Ching Lok, Quinn E Spencer, Hayley C. McCausland, Ciera C. Martinez, and Michael B. Eisen, "Robust Manipulation of the Behavior of *Drosophila melanogaster* by a Fungal Pathogen in the Laboratory," *eLife* 7 (July 26, 2018), https://doi.org/10.7554/elife.34414.
10. Jolet De Ruiter, Sif Arnbjerg-Nielsen, Pascal Herren, Freja Høier, Henrik H. de Fine Licht, and Kaare H. Jensen, "Fungal Artillery of Zombie Flies: Infectious Spore Dispersal Using a Soft Water Cannon," *Journal of the Royal Society Interface* 16, no. 159 (October 30, 2019): 20190448, https://doi.org/10.1098/rsif.2019.0448.

11. Anders Pape Møller, "A Fungus Infecting Domestic Flies Manipulates Sexual Behaviour of Its Host," *Behavioral Ecology and Sociobiology* 33, no. 6 (December 1993): 403–7, https://doi.org/10.1007/bf00170255.
12. Faculty of Science News, University of Copenhagen, "Zombie Fly Fungus Lures Healthy Male Flies to Mate with Female Corpses," July 15, 2022, https://science.ku.dk/english/press/news/2022/zombie-fly-fungus-lures-healthy-male-flies-to-mate-with-female-corpses/.
13. Naundrup et al., "Pathogenic Fungus Uses Volatiles."
14. Fabrizio Fanti and Janusz Kupryjanowicz, "A New Soldier Beetle from Eocene Baltic Amber," *Acta Palaeontologica Polonica* 62, no. 4 (2017): 785–88, https://doi.org/10.4202/app.00388.2017.
15. Roland Thaxter, "The Entomophthoreae of the United States," *Memoirs of the Boston Society of Natural History* 4, no. 6 (1888), https://archive.org/details/cu31924001725724.
16. C. H. Popenoe and E. G. Smyth, "An Epidemic of Fungous Diseases among Soldier Beetles," *Proceedings of the Entomological Society of Washington* 13–14 (1911).
17. D. M. MacLeod and E. Müller-Kögler, "Entomogenous Fungi: Entomophthora Species with Pear-Shaped to Almost Spherical Conidia (Entomophthorales: Entomophthoraceae)," *Mycologia* 65, no. 4 (July 1973): 823–93, https://doi.org/10.2307/3758521.
18. G. R. Carner, "*Entomophthora lampyridarum*, a Fungal Pathogen of the Soldier Beetle, *Chauliognathus Pennsylvanicus*," *Journal of Invertebrate Pathology* 36, no. 3 (November 1980): 394–98, https://doi.org/10.1016/0022-2011(80)90044-0.
19. Donald C. Steinkraus, Ann E. Hajek, and James K. Liebherr, "Zombie Soldier Beetles: Epizootics in the Goldenrod Soldier Beetle, *Chauliognathus pensylvanicus* (Coleoptera: Cantharidae) Caused by *Eryniopsis lampyridarum* (Entomophthoromycotina: Entomophthoraceae)," *Journal of Invertebrate Pathology* 148 (September 2017): 51–59, https://doi.org/10.1016/j.jip.2017.05.002.
20. Raquel Vilela and Leonel Mendoza, "Human Pathogenic Entomophthorales," *Clinical Microbiology Reviews* 31, no. 4 (October 2018), https://doi.org/10.1128/cmr.00014-18.
21. Kathie T. Hodge, Ann E. Hajek, and Andrii P. Gryganskyi, "The First Entomophthoralean Killing Millipedes, *Arthrophaga myriapodina* n. gen. n. sp., Causes Climbing before Host Death," *Journal of Invertebrate Pathology* 149 (October 2017): 135–40, https://doi.org/10.1016/j.jip.2017.08.011.
22. Thomas Eisner, Maria Eisner, and Melody Siegler, *Secret Weapons: Defenses of Insects, Spiders, Scorpions, and Other Many-Legged Creatures* (Cambridge, MA: Belknap Press of Harvard University Press, 2005).
23. Carolyn Elya, Danylo Lavrentovich, Emily Lee, Cassandra Pasadyn, Jasper Duval, Maya Basak, Valerie Saykina et al., "Neural Mechanisms

of Parasite-Induced Summiting Behavior in 'Zombie' *Drosophila*," *eLife* 12 (May 15, 2023), https://doi.org/10.7554/elife.85410.
24. Rohini Subrahmanyam, " 'The Last of Us,' Fruit Fly Edition," *Harvard Gazette*, April 18, 2023, https://news.harvard.edu/gazette/story/2023/04/how-a-mind-controlling-fungal-parasite-turns-insects-into-zombies/.

Chapter 4. The Sleeper Awakens

1. Angie Macias, "Flying Salt Shakers of Death," Cornell Mushroom Blog, Cornell University, February 19, 2013. https://blog.mycology.cornell.edu/2013/02/19/flying-salt-shakers-of-death/.
2. C. L. Marlatt, *The Periodical Cicada: An Account of* Cicada septendecim, *Its Natural Enemies and the Means of Preventing Its Injury: Together with a Summary of the Distribution of the Different Broods* (Washington, DC: US Department of Agriculture, 1898), https://doi.org/10.5962/bhl.title.109853.
3. Benjamin Banneker, "Memoir of Benjamin Banneker: Read before the Maryland Historical Society, at the Monthly Meeting, May 1, 1845, by John H. B. Latrobe, Esq." (Baltimore: John D. Toy, 1845), https://archive.org/details/memoirbenjaminb00socigoog/page/n18/mode/1up.
4. Academy of Natural Sciences, "July 1st, 1851." *Proceedings of the Academy of Natural Sciences of Philadelphia* 5 (1850): 235, http://www.jstor.org/stable/4058846.
5. "Report of the Botanist," Annual Report on the New York State Museum of Natural History, 1871, https://www.biodiversitylibrary.org/page/35614048#page/58/mode/1up.
6. Richard S. Soper, A. J. Delyzer, and Laurence F. R. Smith, "The Genus *Massospora* Entomopathogenic for Cicadas. Part. II. Biology of *Massospora levispora* and Its Host *Okanagana rimosa*, With Notes on *Massospora cicadina* on the Periodical Cicadas," *Annals of the Entomological Society of America* 69, no. 1 (January 15, 1976): 89–95, https://doi.org/10.1093/aesa/69.1.89.
7. Joann White, Phillip Ganter, Richard McFarland, Nancy L. Stanton, and Monte Lloyd, "Spontaneous, Field Tested and Tethered Flight in Healthy and Infected *Magicicada Septendecim* L.," *Oecologia* 57, no. 3 (March 1983): 281–86, https://doi.org/10.1007/bf00377168.
8. John R. Cooley, David C. Marshall, and Kathy B. R. Hill, "A Specialized Fungal Parasite (*Massospora cicadina*) Hijacks the Sexual Signals of Periodical Cicadas (Hemiptera: Cicadidae: *Magicicada*)," *Scientific Reports* 8, no. 1 (January 23, 2018), https://doi.org/10.1038/s41598-018-19813-0.
9. Greg R. Boyce, Emile Gluck-Thaler, Jason C. Slot, Jason Stajich, William J. Davis, Timothy Y. James, John R. Cooley et al., "Psychoactive Plant- and Mushroom-associated Alkaloids from Two Behavior

Modifying Cicada Pathogens," *Fungal Ecology* 41 (October 2019): 147–64, https://doi.org/10.1016/j.funeco.2019.06.002.

10. Ayan Ahmed, Manuel J. Ruiz, Kathrin Cohen Kadosh, Robert Patton, and Davinia M. Resurrección, "Khat and Neurobehavioral Functions: A Systematic Review," *PLOS ONE* 16, no. 6 (June 10, 2021): e0252900, https://doi.org/10.1371/journal.pone.0252900.
11. Hans-Rudolf Frischknecht and Peter G. Waser, "Actions of Hallucinogens on Ants (*Formica pratensis*)—II. Effects of Amphetamine, LSD and Delta-9-tetrahydrocannabinol," *General Pharmacology: The Vascular System* 9, no. 5 (1978): 375–80, https://doi.org/10.1016/0306-3623(78)90078-2.
12. Gary L. Brookhart, Robert S. Edgecomb, and Larry L. Murdock, "Amphetamine and Reserpine Deplete Brain Biogenic Amines and Alter Blow Fly Feeding Behavior," *Journal of Neurochemistry* 48, no. 4 (April 1987): 1307–15, https://doi.org/10.1111/j.1471-4159.1987.tb05662.x.
13. Gene Kritsky, "One for the Books: The 2021 Emergence of the Periodical Cicada Brood X," *American Entomologist* 67, no. 4 (December 9, 2021): 40–46, https://doi.org/10.1093/ae/tmab059.
14. Linda Gaudino, "Eating Cicadas? Brood X Emergence Inspires NJ Students to Take a Bite," NBC New York, September 17, 2021, https://www.nbcnewyork.com/news/local/eating-cicadas-brood-x-emergence-inspires-nj-students-to-take-a-bite/3096546/.
15. Mike McKinnon (@mikamckinnon), "Topical once again: Licking rocks is common field practice as taste & texture can be diagnostic, but not every rock is safe to lick. A thread on technique, pop culture, tasting notes, & Rocks To Not Lick," Twitter/X, January 11, 2019, 5:46 p.m. ET, https://twitter.com/mikamckinnon/status/1083857766148657154?lang=en.

Chapter 5. Attack of the Ooze

1. Isabel Reche, G. D'Orta, Natalie Mladenov, Danielle M. Winget, and Curtis A. Suttle, "Deposition Rates of Viruses and Bacteria above the Atmospheric Boundary Layer," *ISME Journal* 12, no. 4 (April 2018): 1154–62, https://doi.org/10.1038/s41396-017-0042-4.
2. Ravi P. Subramanian, Julia H. Wildschutte, Crystal Russo, and John M. Coffin, "Identification, Characterization, and Comparative Genomic Distribution of the HERV-K (HML-2) Group of Human Endogenous Retroviruses," *Retrovirology* 8 (November 2011), https://doi.org/10.1186/1742-4690-8-90.
3. Luis F. Camarillo-Guerrero, Alexandre Almeida, Guillermo Rangel-Piñeros, Robert D. Finn, and Trevor D. Lawley, "Massive Expansion of Human Gut Bacteriophage Diversity," *Cell* 184, no. 4 (February 18, 2021): 1098–109.e9, https://doi.org/10.1016/j.cell.2021.01.029.

4. Dennis Carroll, Peter Daszak, Nathan Wolfe, George F. Gao, Carlos Médicis Morel, S.P. Morzaria, Ariel Pablos-Méndez et al., "The Global Virome Project," *Science* 359, no. 6378 (February 23, 2018): 872–74, https://doi.org/10.1126/science.aap7463.
5. "Microbiology by Numbers," *Nature Reviews Microbiology* 9 (2011): 628, https://doi.org/10.1038/nrmicro2644.
6. Neil Athey, "Virus Straight from a 'Zombie Horror Film' Sending Lancashire Caterpillars on a Relentless March to Their Death," *Lancashire Telegraph*, July 31, 2017, https://www.lancashiretelegraph.co.uk/news/15444431.virus-straight-zombie-horror-film-sending-lancashire-caterpillars-relentless-march-death/.
7. Dave Goulson, "*Wipfelkrankheit*: Modification of Host Behaviour during Baculoviral Infection," *Oecologia* 109, no. 2 (January 10, 1997): 219–28, https://doi.org/10.1007/s004420050076.
8. Manli Wang, "Cross-Talking between Baculoviruses and Host Insects towards a Successful Infection," *Philosophical Transactions of the Royal Society B* 374, no. 1767 (January 14, 2019): 20180324, https://doi.org/10.1098/rstb.2018.0324.
9. Joseph S. Elkinton and Andrew M. Liebhold, "Population Dynamics of Gypsy Moth in North America," *Annual Review of Entomology* 35 (January 1990): 571–96, https://doi.org/10.1146/annurev.en.35.010190.003035.
10. Rosalind R. James and Zengzhi Li, "From Silkworms to Bees," in *Insect Pathology*, 2nd ed., ed. Fernando E. Vega and Harry K. Kaya (Elsevier eBooks, 2012): 425–59, https://doi.org/10.1016/b978-0-12-384984-7.00012-9.
11. Shizuo G. Kamita, Koukichi Nagasaka, Josie W. Chua, Toru Shimada, Kazuei Mita, Masahiko Kobayashi, Susumu Maeda et al., "A Baculovirus-Encoded Protein Tyrosine Phosphatase Gene Induces Enhanced Locomotory Activity in a Lepidopteran Host," *Proceedings of the National Academy of Sciences* 102, no. 7 (February 7, 2005): 2584–89, https://doi.org/10.1073/pnas.0409457102.
12. Kelli Hoover, Michael Grove, Matthew R. Gardner, David Hughes, James McNeil, and James M. Slavicek, "A Gene for an Extended Phenotype," *Science* 333, no. 6048 (September 9, 2011): 1401, https://doi.org/10.1126/science.1209199.
13. Yue Han, Stineke van Houte, Monique M. van Oers, and Vera I. D. Ros, "Timely Trigger of Caterpillar Zombie Behaviour: Temporal Requirements for Light in Baculovirus-Induced Tree-top Disease," *Parasitology* 145, no. 6 (November 16, 2017): 822–27, https://doi.org/10.1017/s0031182017001822.
14. X. Liu, Zhiqiang Tian, Limei Cai, Zhizhang Shen, J. P. Michaud, Lin Zhu, Shuo Yan, et al., "Baculoviruses Hijack the Visual Perception of Their Caterpillar Hosts to Induce Climbing Behaviour Thus Promoting

Virus Dispersal," *Molecular Ecology* 31, no. 9 (May 2022): 2752–65, https://doi.org/10.1111/mec.16425.

15. Stineke van Houte, Monique M. Van Oers, Yue Han, Just M. Vlak, and V.I.D. Ros, "Baculovirus Infection Triggers a Positive Phototactic Response in Caterpillars: A Response to Dobson et al. (2015)," *Biology Letters* 11, no. 10 (October 2015), https://doi.org/10.1098/rsbl.2015.0633.
16. J. D. Podgwaite and P. R. Galipeau, "Effect of Nucleopolyhedrosis Virus on Two Avian Predators of the Gypsy Moth," Research Note NE-251 (Broomall, PA: US Forest Service Research and Development, 1978), https://www.fs.usda.gov/research/treesearch/19233.
17. Dulce Rebolledo, Rodrigo Lasa, Roger Guevara, Rosa Murillo, and Trevor Williams, "Baculovirus-Induced Climbing Behavior Favors Intraspecific Necrophagy and Efficient Disease Transmission in *Spodoptera exigua*," *PLOS ONE* 10, no. 9 (September 24, 2015): e0136742, https://doi.org/10.1371/journal.pone.0136742.
18. John L. Capinera, "Cabbage Looper, *Trichoplusia ni* (Hübner) (Insecta: Lepidoptera: Noctuidae)," Ask IFAS, University of Florida, February 20, 2024, https://edis.ifas.ufl.edu/publication/IN273.
19. A. M. Heimpel, E. D. Thomas, Jean R. Adams, and Lee Smith, "The Presence of Nuclear Polyhedrosis Viruses of *Trichoplusia ni* on Cabbage from the Market Shelf," *Environmental Entomology* 2, no. 1 (February 1, 1973): 72–75, https://doi.org/10.1093/ee/2.1.72.
20. Joel Henrique Ellwanger and José Artur Bogo Chies, "Zoonotic Spillover: Understanding Basic Aspects for Better Prevention," *Genetics and Molecular Biology* 44, no. 1 supplement 1 (2021), https://doi.org/10.1590/1678-4685-gmb-2020-0355.
21. Elisabeth A. Herniou, Julie A. Olszewski, David R. O'Reilly, and Jenny S. Cory, "Ancient Coevolution of Baculoviruses and Their Insect Hosts," *Journal of Virology* 78, no. 7 (April 1, 2004): 3244–51, https://doi.org/10.1128/jvi.78.7.3244-3251.2004.

Chapter 6. The Original Chest-Bursters

1. Justin Chang and Glenn Whipp, "Our Critics Discuss Ridley Scott's Sci-fi Horror Classic *Alien*," *Los Angeles Times*, May 28, 2020, https://www.latimes.com/entertainment-arts/movies/story/2020-05-27/ultimate-summer-movie-showdown-alien.
2. Philippa Brewer, Kaitlyn Burton, Adrian Lister, Amy Scott-Murray, Jennifer Pular, Lorna Steel, Emily Keeble et al., "*Toxodon platensis*—Cranium," *Darwin's Fossil Mammals*, Natural History Museum Data Portal, 2018, https://data.nhm.ac.uk/dataset/darwins-fossil-mammals/resource/dadf577f-fa52-4f48-8cdb-d11c98448009.
3. Charles Darwin, Letter to Asa Gray, May 22, 1860, Darwin Correspondence Project, https://www.darwinproject.ac.uk/letter/DCP-LETT-2814.xml.

4. Francesco Pennacchio and Michael R. Strand, "Evolution of Developmental Strategies in Parasitic Hymenoptera," *Annual Review of Entomology* 51 (January 2006): 233–58, https://doi.org/10.1146/annurev.ento.51.110104.151029.
5. Seraina Klopfstein, Bernardo F. Santos, Mark R. Shaw, Mabel Alvarado, Andrew M. R. Bennett, Davide Dal Pos, Madalene Giannotta et al., "Darwin Wasps: A New Name Heralds Renewed Efforts to Unravel the Evolutionary History of Ichneumonidae," *Entomological Communications* 1 (December 8, 2019): ec01006, https://doi.org/10.37486/2675-1305.ec01006.
6. Mirko Wölfling and Michael Rostás, "Parasitoids Use Chemical Footprints to Track Down Caterpillars," *Communicative & Integrative Biology* 2, no. 4 (July 1, 2009): 353–55, https://doi.org/10.4161/cib.2.4.8612.
7. Consuelo M. De Moraes, W. J. Lewis, Paul W. Paré, Hans T. Alborn, and James H. Tumlinson, "Herbivore-Infested Plants Selectively Attract Parasitoids," *Nature* 393, no. 6685 (June 1, 1998): 570–73, https://doi.org/10.1038/31219.
8. Jan Suszkiw, "Plants Send SOS When Caterpillars Bite," USDA, *Agricultural Research* 46, no. 10 (October 1998): 20, https://agresearchmag.ars.usda.gov/1998/oct/sos/.
9. Annie Bézier, Faustine Louis, Séverine Jancek, Georges Périquet, Julien Thézé, Gàbor Gyapay, Karine Musset et al., "Functional Endogenous Viral Elements in the Genome of the Parasitoid Wasp *Cotesia congregata*: Insights into the Evolutionary Dynamics of Bracoviruses," *Philosophical Transactions of the Royal Society B* 368, no. 1626 (September 19, 2013): 20130047, https://doi.org/10.1098/rstb.2013.0047.
10. Michael R. Straind and Gaelen R. Burke, "Polydnaviruses: From Discovery to Current Insights," *Virology* 479–480 (May 2015): 393–402, https://doi.org/10.1016/j.virol.2015.01.018.
11. Nancy E. Beckage, Frances F. Tan, Kathleen W. Schleifer, Roni D. Lane, and Lisa L. Cherubin, "Characterization and Biological Effects of *Cotesia congregata* Polydnavirus on Host Larvae of the Tobacco Hornworm, *Manduca sexta*," Archives of *Insect Biochemistry and Physiology* 26, no. 2–3 (January 1, 1994): 165–95, https://doi.org/10.1002/arch.940260209.
12. Catherine Dupuy, Elisabeth Huguet, and Jean-Michel Drezen, "Unfolding the Evolutionary Story of Polydnaviruses," *Virus Research* 117, no. 1 (April 2006): 81–89, https://doi.org/10.1016/j.virusres.2006.01.001.
13. B. B. Fulton, "The Hornworm Parasite, *Apanteles congregatus* (Say) and the Hyperparasite, *Hypopteromalus tabacum* (Fitch)," *Annals of the Entomological Society of America* 33, no. 2 (June 1, 1940): 231–44, https://doi.org/10.1093/aesa/33.2.231.
14. Shelley A. Adamo, Charles E. Linn, and Nancy E. Beckage, "Correlation between Changes in Host Behaviour and Octopamine Levels in the Tobacco Hornworm *Manduca sexta* Parasitized by the Gregarious

Braconid Parasitoid Wasp *Cotesia congregata*," *Journal of Experimental Biology* 200, no. 1 (January 1, 1997): 117–27, https://doi.org/10.1242/jeb.200.1.117.

15. Mitchell S. Dushay and Nancy E. Beckage, "Dose-dependent Separation of *Cotesia congregata*-associated Polydnavirus Effects on *Manduca sexta* Larval Development and Immunity," *Journal of Insect Physiology* 39, no. 12 (December 1993): 1029–40, https://doi.org/10.1016/0022-1910(93)90127-d.
16. Shelley A. Adamo, "Feeding Suppression in the Tobacco Hornworm, *Manduca sexta*: Costs and Benefits to the Parasitic Wasp *Cotesia congregata*," *Canadian Journal of Zoology* 76, no. 9 (September 1998): 1634–40, https://doi.org/10.1139/z98-105.
17. Shelley A. Adamo, Ilya Kovalko, Kurtis F. Turnbull, Russell H. Easy, and Carol I. Miles, "The Parasitic Wasp *Cotesia congregata* Uses Multiple Mechanisms to Control Host (*Manduca sexta*) Behaviour," *Journal of Experimental Biology* 219, no. 23 (December 2016), https://doi.org/10.1242/jeb.145300.
18. Jacques Brodeur and Louise E. M. Vet, "Usurpation of Host Behaviour by a Parasitic Wasp," *Animal Behaviour* 48, no. 1 (July 1994): 187–92, https://doi.org/10.1006/anbe.1994.1225.
19. A. H. Grosman, Arne Janssen, Elaine F. De Brito, Eduardo G. Cordeiro, Felipe Colares, Juliana Oliveira Fonseca, Eraldo Lima et al., "Parasitoid Increases Survival of Its Pupae by Inducing Hosts to Fight Predators," *PLOS ONE* 3, no. 6 (June 4, 2008): e2276, https://doi.org/10.1371/journal.pone.0002276.
20. University of Oxford and Wytham Woods Genome Acquisition Lab, Gavin R. Broad, Darwin Tree of Life Barcoding collective, Wellcome Sanger Institute Tree of Life programme, Wellcome Sanger Institute Scientific Operations: DNA Pipelines, Tree of Life Core Informatics collective, and Darwin Tree of Life Consortium, "The Genome Sequence of a Parasitoid Wasp, *Ichneumon xanthorius* Forster, 1771," Wellcome Open Research 7 (February 10, 2022): 47, https://doi.org/10.12688/wellcomeopenres.17683.1.

Chapter 7. 'Til Death Do Us Part

1. Mike Wall, "Venomous Spiders Spin Weightless Webs in Space," Space.Com, May 28, 2011, https://www.space.com/11818-space-spiders-weightless-webs-station-shuttle.html.
2. William Eberhard, "The Natural History and Behavior of *Hymenoepimecis argyraphaga* (Hymenoptera: Ichneumonidae) a Parasitoid of *Plesiometa argyra* (Araneae: Tetragnathidae)," *Journal of Hymenoptera Research* 9, no. 2 (2000): 220–40, https://biostor.org/reference/271.
3. William G. Eberhard, "Spider Manipulation by a Wasp Larva," *Nature* 406, no. 6793 (July 20, 2000): 255–56, https://doi.org/10.1038/35018636.

4. Philippe Fernandez-Fournier, Samantha Straus, Ruth Victoria Sharpe, and Leticia Avilés, "Behavioural Modification of a Social Spider by a Parasitoid Wasp," *Ecological Entomology* 44, no. 2 (November 4, 2018): 157–62, https://doi.org/10.1111/een.12698.
5. William G. Eberhard, "New Types of Behavioral Manipulation of Host Spiders by a Parasitoid Wasp," *Psyche* 2010, no. 1 (January 2010): 950614, https://doi.org/10.1155/2010/950614.
6. Lui Xia Lee, "Newly Discovered Wasp Turns Social Spiders into Zombies," University of British Columbia News, November 27, 2018, https://news.ubc.ca/2018/11/27/newly-discovered-wasp-turns-social-spiders-into-zombies/.
7. Keizo Takasuka, Tomoki Yasui, Toru Ishigami, Kensuke Nakata, Rikio Matsumoto, Kenichi Ikeda, and Kaoru Maetô, "Host Manipulation by an Ichneumonid Spider Ectoparasitoid That Takes Advantage of Preprogrammed Web-Building Behaviour for Its Cocoon Protection," *Journal of Experimental Biology* 218, no. 15 (August 2015): 2326–32, https://doi.org/10.1242/jeb.122739.
8. Thiago Gechel Kloss, Marcelo O. Gonzaga, Leandro Licursi De Oliveira, and Carlos Frankl Sperber, "Proximate Mechanism of Behavioral Manipulation of an Orb-Weaver Spider Host by a Parasitoid Wasp," *PLOS ONE* 12, no. 2 (February 3, 2017): e0171336, https://doi.org/10.1371/journal.pone.0171336.
9. William G. Eberhard, and Marcelo O. Gonzaga, "Evidence That *Polysphincta*-Group Wasps (Hymenoptera: Ichneumonidae) Use Ecdysteroids to Manipulate the Web-Construction Behaviour of Their Spider Hosts," *Biological Journal of the Linnean Society* 127, no. 2 (June 2019): 429–71, https://doi.org/10.1093/biolinnean/blz044.
10. Gal Haspel and Frederic Libersat, "Wasp Manipulates Cockroach Behavior by Injecting Venom Cocktail into Prey Central Nervous System," *Acta Biologica Hungarica* 55, nos. 1–4 (May 1, 2004): 103–12, https://doi.org/10.1556/abiol.55.2004.1-4.12.
11. Ryan Arvidson, Victor Landa, Sarah Frankenberg, and Michael E. Adams, "Life History of the Emerald Jewel Wasp *Ampulex compressa*," *Journal of Hymenoptera Research* 63 (April 30, 2018): 1–13, https://doi.org/10.3897/jhr.63.21762.
12. Gal Haspel, Lior Rosenberg, and Frederic Libersat, "Direct Injection of Venom by a Predatory Wasp into Cockroach Brain," *Journal of Neurobiology* 56, no. 3 (May 20, 2003): 287–92, https://doi.org/10.1002/neu.10238.
13. Ram Gal, Lior Rosenberg, and Frederic Libersat, "Parasitoid Wasp Uses a Venom Cocktail Injected into the Brain to Manipulate the Behavior and Metabolism of Its Cockroach Prey," *Archives of Insect Biochemistry and Physiology* 60, no. 4 (November 22, 2005): 198–208, https://doi.org/10.1002/arch.20092.

14. Ram Gal and Frederic Libersat, "A Wasp Manipulates Neuronal Activity in the Sub-Esophageal Ganglion to Decrease the Drive for Walking in Its Cockroach Prey," *PLOS ONE* 5, no. 4 (April 7, 2010): e10019, https://doi.org/10.1371/journal.pone.0010019.
15. F. X. Williams, "*Ampulex compressa* (Fabr.): A Cockroach-Hunting Wasp Introduced from New Caledonia into Hawaii," 1942, https://scholarspace.manoa.hawaii.edu/items/7d4d17bd-3c28-4485-bb7a-282812204814.
16. Kenneth C. Catania, "How Not to Be Turned into a Zombie," *Brain, Behavior and Evolution* 92, nos. 1–2 (December 2018): 32–46, https://doi.org/10.1159/000490341.
17. Piotr Ceryngier, Kamila W. Franz, and Jerzy Romanowski, "Distribution, Host Range and Host Preferences of *Dinocampus coccinellae* (Hymenoptera: Braconidae): A Worldwide Database," *European Journal of Entomology* 120 (February 2, 2023): 26–34, https://doi.org/10.14411/eje.2023.004.
18. Fanny Maure, Jacques Brodeur, Nicolas Ponlet, Josée Doyon, Annabelle Firlej, Éric Elguero, and Frédéric Thomas, "The Cost of a Bodyguard," *Biology Letters* 7, no. 6 (June 22, 2011): 843–46, https://doi.org/10.1098/rsbl.2011.0415.
19. Nolwenn M. Dheilly, Fanny Maure, Marc Ravallec, Richard Galinier, Josée Doyon, David Duval, Lucas Léger et al., "Who Is the Puppet Master? Replication of a Parasitic Wasp-Associated Virus Correlates with Host Behaviour Manipulation," *Proceedings of the Royal Society B: Biological Sciences* 282, no. 1803 (March 22, 2015): 20142773, https://doi.org/10.1098/rspb.2014.2773.
20. Nicholas Weiler, "Wasp Virus Turns Ladybugs into Zombie Babysitters," *Science*, February 10, 2015, https://www.science.org/content/article/wasp-virus-turns-ladybugs-zombie-babysitters.
21. Tiana Gayton (@tiana_thebuglady), "Don't give up, Stay strong, Spread your wings & Fly," TikTok, May 11, 2021, https://www.tiktok.com/@tiana_thebuglady/video/6960940631074114822.
22. Erinn P. Fagan-Jeffries, Steven J. Cooper, and Andrew D. Austin, "Three New Species of *Dolichogenidea* Viereck (Hymenoptera, Braconidae, Microgastrinae) from Australia with Exceptionally Long Ovipositors," *Journal of Hymenoptera Research* 64 (June 25, 2018): 177–90, https://doi.org/10.3897/jhr.64.25219.
23. "Newly Discovered Xenomorph Wasp Has Alien-like Lifecycle," University of Adelaide News & Events, June 27, 2018, https://www.adelaide.edu.au/news/news100882.html.
24. Dennis McLellan, "Dan O'Bannon Dies at 63; Screenwriter of *Alien*," *Los Angeles Times*, December 19, 2009, https://www.latimes.com/local/obituaries/la-me-dan-obannon19-2009dec19-story.html.

25. Bruce Weber, "Dan O'Bannon, 63, Who Wrote Screenplay for 'Alien,' Is Dead," *New York Times*, December 20, 2009, https://www.nytimes.com/2009/12/21/movies/21obannon.html.
26. Jason Zinoman, *Shock Value: How a Few Eccentric Outsiders Gave Us Nightmares, Conquered Hollywood, and Invented Modern Horror* (Penguin, 2012).
27. Josh Weiss, "James Cameron Explains How a Stinging Nightmare Inspired the Xenomorph Queen Climax in *Aliens*," SYFY, December 13, 2021, https://www.syfy.com/syfy-wire/james-cameron-aliens-xenomorph-queen-wasp-dream.

Chapter 8. Ant Decapitators and the Rise of the ZomBees

1. "Order Diptera (Flies)," Description of Order and Families in British Columbia (website), University of British Columbia, Department of Zoology, accessed July 20, 2024, https://www.zoology.ubc.ca/bcdiptera/Order%20Diptera%20Text%20Files/family_descriptions.htm.
2. Brian V. Brown, "Small Size No Protection for Acrobat Ants: World's Smallest Fly Is a Parasitic Phorid (Diptera: Phoridae)," *Annals of the Entomological Society of America* 105, no. 4 (July 1, 2012): 550–54, https://doi.org/10.1603/an12011.
3. Brian V. Brown, "A Second Contender for 'World's Smallest Fly' (Diptera: Phoridae)," *Biodiversity Data Journal* 6 (January 24, 2018): e22396, https://doi.org/10.3897/bdj.6.e22396.
4. Phorid.net, https://phorid.net/.
5. Lloyd Morrison, "Biology of *Pseudacteon* (Diptera: Phoridae) Ant Parasitoids and Their Potential to Control Imported *Solenopsis* Fire Ants (Hymenoptera: Formicidae)," *Recent Research Developments in Entomology* 3 (2000), 1–13.
6. Lloyd Morrison, "*Pseudacteon* spp. (Diptera: Phoridae)," Biological Control: A Guide to Natural Enemies in North America (website), Cornell University, College of Agriculture and Life Sciences, https://biocontrol.entomology.cornell.edu/parasitoids/pseudacteon.php.
7. Adilson Ariza Zacaro and Sanford D. Porter, "Female Reproductive System of the Decapitating Fly *Pseudacteon Wasmanni* Schmitz (Diptera: Phoridae)," *Arthropod Structure & Development* 31, no. 4 (April 2003): 329–37, https://doi.org/10.1016/s1467-8039(02)00049-x.
8. Sanford Porter, "Biology and Behavior of *Pseudacteon* Decapitating Flies (Diptera: Phoridae) That Parasitize *Solenopsis* Fire Ants (Hymenoptera: Formicidae)," *Florida Entomologist* 81, no. 3 (September 1998): 292–309, https://journals.flvc.org/flaent/article/view/74833/72491.
9. Li Chen and Sanford D. Porter, "Biology of *Pseudacteon* Decapitating Flies (Diptera: Phoridae) That Parasitize Ants of the *Solenopsis Saevissima* Complex (Hymenoptera: Formicidae) in South America," *Insects* 11, no. 2 (February 6, 2020): 107, https://doi.org/10.3390/insects11020107.

10. J. T. King, Jean R. Starkey, Valerie Renee Holmes, Robert T. Puckett, and Edward L. Vargo, "Quit Bugging Me: Phorid Fly Parasitoids Affect Expression of an Immune Gene in Foraging Fire Ant Workers," *Insectes Sociaux* 70, no. 3 (August 24, 2023): 339–51, https://doi.org/10.1007/s00040-023-00930-7.
11. Seth Johnson, Don Henne, Anna Mészáros, and Lee Eisenberg, "Zombie Fire Ants: Biological Control of the Red Imported Fire Ant in Louisiana with Decapitating Phorid Flies," *Louisiana Agriculture*, Fall 2010, https://www.lsuagcenter.com/portals/communications/publications/agmag/archive/2010/fall/zombiefireantsbiologicalcontroloftheredimportedfireant-inlouisianawithdecapitatingphori.
12. Donald C. Henne and S. J. Johnson, "Zombie Fire Ant Workers: Behavior Controlled by Decapitating Fly Parasitoids," *Insectes Sociaux* 54, no. 2 (March 12, 2007): 150–53, https://doi.org/10.1007/s00040-007-0924-y.
13. "Fire Ant Bites," Cleveland Clinic, June 28, 2022, https://my.clevelandclinic.org/health/diseases/23362-fire-ant-bites.
14. Mattia Menchetti, Enrico Schifani, Antonio Alicata, Laura Cardador, Elisabetta Sbrega, Eric Toro-Delgado, and Roger Vila, "The Invasive Ant *Solenopsis invicta* Is Established in Europe," *Current Biology* 33, no. 17 (September 11, 2023): R896–97, https://doi.org/10.1016/j.cub.2023.07.036.
15. "Fire Ants," Fort Smith National Historic Site, US National Park Service, accessed June 18, 2024, https://www.nps.gov/fosm/learn/nature/fire-ants.htm.
16. "Fire Ants and Other Burning Problems," USDA Tellus, accessed June 18, 2024, https://tellus.ars.usda.gov/stories/articles/fire-ants-and-other-burning-problems.
17. Jill Lee, "Ouch! The Fire Ant Saga Continues," USDA *AgResearch Magazine* 47, no. 9 (September 1999), https://agresearchmag.ars.usda.gov/1999/sep/anti/.
18. Robert T. Puckett and Marvin K. Harris, "Phorid Flies, *Pseudacteon* spp. (Diptera: Phoridae), Affect Forager Size Ratios of Red Imported Fire Ants *Solenopsis invicta* (Hymenoptera: Formicidae) in Texas," *Environmental Entomology* 39, no. 5 (October 1, 2010): 1593–1600, https://doi.org/10.1603/en09189.
19. Li Chen and Sanford D. Porter, "Biology of *Pseudacteon* Decapitating Flies (Diptera: Phoridae) That Parasitize Ants of the *Solenopsis Saevissima* Complex (Hymenoptera: Formicidae) in South America," *Insects* 11, no. 2 (February 6, 2020): 107, https://doi.org/10.3390/insects11020107.
20. Patricia J. Folgarait, Robert M. Plowes, Carolina Gomila, and Lawrence E. Gilbert, "A Small Parasitoid of Fire Ants, *Pseudacteon obtusitus* (Diptera: Phoridae): Native Range Ecology and Laboratory Rearing," *Florida Entomologist* 103, no. 1 (April 8, 2020): 9–15, https://doi.org/10.1653/024.103.0402.

21. Li Chen and Henry Y. Fadamiro, "*Pseudacteon* Phorid Flies: Host Specificity and Impacts on *Solenopsis* Fire Ants," *Annual Review of Entomology* 63 (January 7, 2018): 47–67, https://doi.org/10.1146/annurev-ento-020117-043049.
22. Li Chen and Lloyd W. Morrison, "Importation Biological Control of Invasive Fire Ants with Parasitoid Phorid Flies—Progress and Prospects," *Biological Control* 154 (March 1, 2021): 104509, https://doi.org/10.1016/j.biocontrol.2020.104509.
23. "Flight of the Living Dead: Dr. John Hafernik at TEDxJacksonHole," TEDx Talks, video, 12:15, October 31, 2012, https://www.youtube.com/watch?v=BWl_1vrSxMc.
24. Andrew Core, Charles Runckel, Jonathan Ivers, Christopher Quock, Travis Siapno, Seraphina DeNault, Brian V. Brown et al., "A New Threat to Honey Bees, the Parasitic Phorid Fly *Apocephalus borealis*," *PLOS ONE* 7, no. 1 (January 3, 2012): e29639, https://doi.org/10.1371/journal.pone.0029639.
25. "Deadly Fly Parasite Spotted for First Time in Honey Bees," EurekAlert!, AAAS, January 3, 2012, https://www.eurekalert.org/news-releases/792897.
26. " 'ZomBee' Hunters Needed," *SF State News*, August 2013, https://news.sfsu.edu/archive/zombee-hunters-needed.html.
27. Erik Tihelka, John E. Hafernik, Brian V. Brown, Christopher Quock, Andrew G. Zink, Sofia Croppi, Chenyang Cai et al., "Global Invasion Risk of *Apocephalus borealis*, a Honey Bee Parasitoid," *Apidologie* 52, no. 6 (December 21, 2021): 1128–40, https://doi.org/10.1007/s13592-021-00892-4.
28. Brian V. Brown, "Not Just Honey Bees and Bumble Bees: First Record of 'Zombie' Flies (Diptera: Phoridae) from a Carpenter Bee (Hymenoptera: Apidae: Xylocopinae)," *Pan-Pacific Entomologist* 93, no. 3 (October 9, 2017): 113–14, https://doi.org/10.3956/2017-93.2.113.
29. Tarli E. Conroy and Luke Holman, "Social Immunity in the Honey Bee: Do Immune-Challenged Workers Enter Enforced or Self-imposed Exile?," *Behavioral Ecology and Sociobiology* 76 (February 11, 2022): 32, https://doi.org/10.1007/s00265-022-03139-z.

Chapter 9. What Lies Within

1. Gerald Thorne, "The Hairworm, *Gordius robustus* Leidy, as a Parasite of the Mormon Cricket, *Anabrus simplex* Haldeman," *Journal of the Washington Academy of Sciences* 30, no. 5 (1940): 219–31, http://www.jstor.org/stable/24529727.
2. Andreas Schmidt-Rhaesa and Reinhard Ehrmann, "Horsehair Worms (Nematomorpha) as Parasites of Praying Mantids with a Discussion of Their Life Cycle," *Zoologischer Anzeiger* 240, no. 2 (2001): 167–79, https://doi.org/10.1078/0044-5231-00014.

3. "These Hairworms Eat a Cricket Alive and Control Its Mind," Deep Look, video, 4:35, February 12, 2019, https://www.youtube.com/watch?v=YB6O7jS_VBM.
4. Frédéric Thomas, Andreas Schmidt-Rhaesa, Guillaume Martin, C. Manu, Patrick Durand, and François Renaud, "Do Hairworms (Nematomorpha) Manipulate the Water Seeking Behaviour of Their Terrestrial Hosts?," *Journal of Evolutionary Biology* 15, no. 3 (May 1, 2002): 356–61, https://doi.org/10.1046/j.1420-9101.2002.00410.x.
5. Fleur Ponton, Camille Lebarbenchon, Thierry Léfèvre, David G. Biron, David Duneau, David Hughes, and Frédéric Thomas, "Parasite Survives Predation on Its Host," *Nature* 440 (April 6, 2006): 756, https://doi.org/10.1038/440756a.
6. David G. Biron, Laurent Marché, Fleur Ponton, Hugh D. Loxdale, Nathalie Galéotti, L. Renault, Cécile Joly et al., "Behavioural Manipulation in a Grasshopper Harbouring Hairworm: A Proteomics Approach," *Proceedings of the Royal Society B: Biological Sciences* 272, no. 1577 (August 31, 2005): 2117–26, https://doi.org/10.1098/rspb.2005.3213.
7. Nasono Obayashi, Yasushi Iwatani, Midori Sakura, Satoshi Tamotsu, Ming-Chung Chiu, and Tadanobu Sato, "Enhanced Polarotaxis Can Explain Water-Entry Behaviour of Mantids Infected with Nematomorph Parasites," *Current Biology* 31, no. 12 (June 21, 2021): R777–78, https://doi.org/10.1016/j.cub.2021.05.001.
8. Fleur Ponton, Fernando Otálora-Luna, Thierry Lefèvre, Patrick M. Guerin, Camille Lebarbenchon, David Duneau, David G. Biron, et al., "Water-Seeking Behavior in Worm-Infected Crickets and Reversibility of Parasitic Manipulation," *Behavioral Ecology* 22, no. 2 (March–April 2011): 392–400, https://doi.org/10.1093/beheco/arq215.
9. N. A. Cobb, "Nematodes and Their Relationships," in *Yearbook of the United States Department of Agriculture, 1914* (Washington, DC: 1915), 457–90, https://search.nal.usda.gov/permalink/01NAL_INST/27vehl/alma9915491010107426.
10. W. M. Mann, "The Stanford Expedition to Brazil, 1911, John C. Branner, Director: The Ants of Brazil," *Bulletin of the Museum of Comparative Zoology* 60 (1916): 399–490.
11. Stephen P. Yanoviak, Michael Kaspari, Robert Dudley, and George O. Poinar, "Parasite-Induced Fruit Mimicry in a Tropical Canopy Ant," *American Naturalist* 171, no. 4 (April 2008): 536–44, https://doi.org/10.1086/528968.
12. Ernst Zeller, "Ueber Leucochloridium paradoxum Carus und die weitere entwickelung seiner Distomenbrut," *Zeitschrift Für Wissenschaftliche Zoologie* 24, no. 1 (1874), https://www.digitale-sammlungen.de/en/view/bsb11190880?page=574.
13. *The Annals and Magazine of Natural History*, Taylor and Francis (1875), https://www.biodiversitylibrary.org/item/53338#page/9/mode/1up.

14. Wanda Wesołowska and Tomasz Wesołowski, "Do *Leucochloridium* Sporocysts Manipulate the Behaviour of Their Snail Hosts?," *Journal of Zoology* 292, no. 3 (March 2014): 151–55, https://doi.org/10.1111/jzo.12094.
15. W. Hohorst and G. Graefe, "Ameisen: Obligatorische Zwischenwirte des Lanzettegels (*Dicrocoelium dendriticum*)," *Naturwissenschaften* 48 (January 1961): 229–30, https://doi.org/10.1007/bf00597502.
16. Daniel Martín-Vega, Amin Garbout, Farah K. Ahmed, Martina Wicklein, Cameron P. Goater, D. D. Colwell, and M. J. R. Hall, "3D Virtual Histology at the Host/Parasite Interface: Visualisation of the Master Manipulator, *Dicrocoelium dendriticum*, in the Brain of Its Ant Host," *Scientific Reports* 8 (June 5, 2018): 8587, https://doi.org/10.1038/s41598-018-26977-2.

Chapter 10. Our Zombie Future

1. Alvaro Aguilar-Setién, Nidia Aréchiga-Ceballos, Gary A. Balsamo, Amy J. Behrman, Hannah K. Frank, Gary R. Fujimoto, Elizabeth Gilman Duane et al., "Biosafety Practices When Working with Bats: A Guide to Field Research Considerations," *Applied Biosafety* 27, no. 3 (September 2022): 169–90, https://doi.org/10.1089/apb.2022.0019.
2. Louise H. Taylor and Louis H. Nel, "Global Epidemiology of Canine Rabies: Past, Present, and Future Prospects," *Veterinary Medicine: Research and Reports* 6 (November 1, 2015): 361–71, https://doi.org/10.2147/vmrr.s51147.
3. "Rabies," World Health Organization, June 5, 2024, https://www.who.int/news-room/fact-sheets/detail/rabies.
4. Kartsten Hueffer, Shailesh Khatri, Shane Alexander Rideout, Michael B. Harris, Roger L. Papke, Clare Stokes, and Marvin K. Schulte, "Rabies Virus Modifies Host Behaviour through a Snake-Toxin Like Region of Its Glycoprotein That Inhibits Neurotransmitter Receptors in the CNS," *Scientific Reports* 7 (October 9, 2017): 12818, https://doi.org/10.1038/s41598-017-12726-4.
5. "Control of Neglected Tropical Diseases: Control and Elimination Strategies—Rabies," World Health Organization, accessed June 19, 2024, https://www.who.int/teams/control-of-neglected-tropical-diseases/rabies/control-and-elimination-strategies.
6. "US Declared Canine-Rabies Free," CDC Newsroom, September 7, 2007, https://www.cdc.gov/media/pressrel/2007/r070907.htm.
7. "USDA Announces Rabies Action Plan to Protect Public, Domestic Animals and Wildlife," USDA Animal and Plant Health Inspection Service, accessed July 20, 2024, https://www.aphis.usda.gov/news/agency-announcements/usda-announces-rabies-action-plan-protect-public-domestic-animals.
8. Sarah Marshall, "USU Researchers Develop Potential Cure for Rabies Infection," USU News, September 28, 2023, https://news.usuhs.edu

/2023/09/usu-researchers-develop-potential-cure.html; Kate E. Mastraccio, Celeste Huaman, Si'Ana A. Coggins, Caitlyn Clouse, Madeline Rader, Lianying Yan, Pratyusha Mandal et al., "mAb Therapy Controls CNS-Resident Lyssavirus Infection via a CD4 T Cell-Dependent Mechanism," *EMBO Molecular Medicine* 15 (September 28, 2023): e16394, https://doi.org/10.15252/emmm.202216394.

9. Mahbobeh Montazeri, Tahereh Mikaeili Galeh, Mahmood Moosazadeh, Shahabeddin Sarvi, Samira Dodangeh, Javad Javidnia, Mehdi Sharif et al., "The Global Serological Prevalence of *Toxoplasma gondii* in Felids during the Last Five Decades (1967–2017): A Systematic Review and Meta-Analysis," *Parasites & Vectors* 13 (February 17, 2020): 82, https://doi.org/10.1186/s13071-020-3954-1.
10. Rupshi Mitra, Robert M. Sapolsky, and Ajai Vyas, "*Toxoplasma gondii* Infection Induces Dendritic Retraction in Basolateral Amygdala accompanied by Reduced Corticosterone Secretion," *Disease Models & Mechanisms* 6, no. 2 (March 2013): 516–20, https://doi.org/10.1242/dmm.009928.
11. Connor J. Meyer, Kira A. Cassidy, Erin E. Stahler, Ellen E. Brandell, Colby B. Anton, Daniel R. Stahler, and Douglas W. Smith, "Parasitic Infection Increases Risk-Taking in a Social, Intermediate Host Carnivore," *Communications Biology* 5 (November 24, 2022): 1180, https://doi.org/10.1038/s42003-022-04122-0.
12. Madlaina Boillat, Pierre-Mehdi Hammoudi, Sunil Kumar Dogga, Stéphane Pagès, Maged Goubran, Iván Rodríguez, and Dominique Soldati-Favre, "Neuroinflammation-Associated Aspecific Manipulation of Mouse Predator Fear by *Toxoplasma Gondii*," *Cell Reports* 30, no. 2 (January 14, 2020): P320–34.e6, https://doi.org/10.1016/j.celrep.2019.12.019.
13. "New Light Cast on the 'Crazy Cat Lady' Parasite," Scimex, January 15, 2020, https://www.scimex.org/newsfeed/new-light-cast-on-the-crazy-cat-lady-parasite.
14. Eliot Barford, "Parasite Makes Mice Lose Fear of Cats Permanently," *Nature*, September 18, 2013, https://doi.org/10.1038/nature.2013.13777.
15. Eben Gering, Zachary M. Laubach, Patty Sue D. Weber, Gisela Soboll Hussey, Kenna D. S. Lehmann, Tracy M. Montgomery, Julie W. Turner et al., "*Toxoplasma Gondii* Infections Are Associated with Costly Boldness toward Felids in a Wild Host," *Nature Communications* 12 (June 22, 2021): 3842, https://doi.org/10.1038/s41467-021-24092-x.
16. Melissa A. Miller, Michael E. Grigg, Christine Kreuder, E. R. James, Ann C. Melli, Paul R. Crosbie, David A. Jessup et al., "An Unusual Genotype of *Toxoplasma gondii* Is Common in California Sea Otters (*Enhydra lutris nereis*) and Is a Cause of Mortality," *International Journal for Parasitology* 34, no. 3 (March 9, 2004): 275–84, https://doi.org/10.1016/j.ijpara.2003.12.008.

17. "About Toxoplasmosis," US Centers for Disease Control and Prevention, accessed June 20, 2024, https://www.cdc.gov/parasites/toxoplasmosis/index.html.
18. Despina G. Contopoulos-Ioannidis, Maria G Gianniki, Angeline Truong, and José G. Montoya, "Toxoplasmosis and Schizophrenia: A Systematic Review and Meta-Analysis of Prevalence and Associations and Future Directions," *Psychiatric Research and Clinical Practice* 4, no. 2 (Summer 2022): 48–60, https://doi.org/10.1176/appi.prcp.20210041.
19. Lies de Haan, Arjen L. Sutterland, Jasper V. Schotborgh, Frederike Schirmbeck, and Lieuwe de Haan, "Association of *Toxoplasma gondii* Seropositivity with Cognitive Function in Healthy People," *JAMA Psychiatry* 78, no. 10 (October 1, 2021): 1103, https://doi.org/10.1001/jamapsychiatry.2021.1590.
20. Jaroslav Flegr, Jan Havlíček, Petr Kodym, Marek Malý, and Zbyněk Smahel, "Increased Risk of Traffic Accidents in Subjects with Latent Toxoplasmosis: A Retrospective Case-Control Study," *BMC Infectious Diseases* 2 (July 2, 2002): 11, https://doi.org/10.1186/1471-2334-2-11.
21. Guillaume Fond, Delphine Capdevielle, Alexandra Macgregor, J. Attal, A. Larue, Marie Brittner, Déborah Ducasse et al., "*Toxoplasma gondii*: Un Rôle Potentiel dans la Genèse de Troubles Psychiatriques. Une Revue Systématique de la Littérature (*Toxoplasma gondii*: A Potential Role in the Genesis of Psychiatric Disorders. A Systematic Review of the Literature)," *L'Encéphale* 39, no. 1 (February 2013): 38–43, https://doi.org/10.1016/j.encep.2012.06.014.
22. Robin Kopecký, Lenka Příplatová, Silvia Boschetti, Konrad Talmont-Kaminski, and Jaroslav Flegr, "Le Petit Machiavellian Prince: Effects of Latent Toxoplasmosis on Political Beliefs and Values," *Evolutionary Psychology* 20, no. 3 (July 29, 2022): 147470492211126, https://doi.org/10.1177/14747049221112657.
23. Stefanie K. Johnson, Markus Fitza, Daniel Lerner, Dana M. Calhoun, Marissa A. Beldon, Elsa Chan, and Pieter T. J. Johnson, "Risky Business: Linking *Toxoplasma Gondii* Infection and Entrepreneurship Behaviours across Individuals and Countries," *Proceedings of the Royal Society B: Biological Sciences* 285, no. 1883 (July 25, 2018): 20180822, https://doi.org/10.1098/rspb.2018.0822.
24. Trent Knoss, "This Cat-Borne Parasite Might Just Make You More Entrepreneurial," CU Boulder Today, August 6, 2018, https://www.colorado.edu/today/2018/07/25/cat-borne-parasite-might-just-make-you-more-entrepreneurial.
25. Ali S. Khan, "Preparedness 101: Zombie Apocalypse," US Centers for Disease Control and Prevention, Public Health Matters (blog), May 16, 2011, https://web.archive.org/web/20111122033654/http://blogs.cdc.gov/publichealthmatters/2011/05/preparedness-101-zombie-apocalypse/.

26. "Preparedness 101: Zombie Apocalypse." Red Cross Canada (blog), February 1, 2013, https://www.redcross.ca/blog/2013/2/preparedness-101-zombie-apocalypse.
27. "Ready Campaign Promotes Disaster Preparedness with 'Zombieland: Double Tap,'" Federal Emergency Management Agency (FEMA), press release, October 9, 2019, https://www.fema.gov/press-release/20230425/ready-campaign-promotes-disaster-preparedness-zombieland-double-tap.
28. "The True Death Toll of COVID-19: Estimating Global Excess Mortality," World Health Organization, May 20, 2021, https://www.who.int/data/stories/the-true-death-toll-of-covid-19-estimating-global-excess-mortality.
29. Athena Aktipis and Joe Alcock, "A Year into the Pandemic, the Coronavirus Is Messing with Our Minds as Well as Our Bodies," *The Conversation*, March 8, 2021, https://theconversation.com/a-year-into-the-pandemic-the-coronavirus-is-messing-with-our-minds-as-well-as-our-bodies-155213.
30. "Chicago Wouldn't Last Long under Zombie Invasion, Model Finds," Newswise, November 4, 2016, https://www.newswise.com/articles/chicago-wouldn-t-last-long-under-zombie-invasion-model-finds.
31. "Learning from the Undead: Simulating Zombie Plagues in Finland Could Help Slow Down Next Pandemic," October 26, 2023, https://www.newswise.com/articles/learning-from-the-undead-simulating-zombie-plagues-in-finland-could-help-slow-down-next-pandemic.

Index